Function Classes of Cauchy Continuous Maps

MONOGRAPHS AND TEXTBOOKS IN PURE AND APPLIED MATHEMATICS

1. *K. Yano,* Integral Formulas in Riemannian Geometry (1970)*(out of print)*
2. *S. Kobayashi,* Hyperbolic Manifolds and Holomorphic Mappings (1970) *(out of print)*
3. *V. S. Vladimirov,* Equations of Mathematical Physics (A. Jeffrey, editor; A. Littlewood, translator) (1970) *(out of print)*
4. *B. N. Pshenichnyi,* Necessary Conditions for an Extremum (L. Neustadt, translation editor; K. Makowski, translator) (1971)
5. *L. Narici, E. Beckenstein, and G. Bachman,* Functional Analysis and Valuation Theory (1971)
6. *D. S. Passman,* Infinite Group Rings (1971)
7. *L. Dornhoff,* Group Representation Theory (in two parts). Part A: Ordinary Representation Theory. Part B: Modular Representation Theory (1971, 1972)
8. *W. Boothby and G. L. Weiss (eds.),* Symmetric Spaces: Short Courses Presented at Washington University (1972)
9. *Y. Matsushima,* Differentiable Manifolds (E. T. Kobayashi, translator) (1972)
10. *L. E. Ward, Jr.,* Topology: An Outline for a First Course (1972) *(out of print)*
11. *A. Babakhanian,* Cohomological Methods in Group Theory (1972)
12. *R. Gilmer,* Multiplicative Ideal Theory (1972)
13. *J. Yeh,* Stochastic Processes and the Wiener Integral (1973) *(out of print)*
14. *J. Barros-Neto,* Introduction to the Theory of Distributions (1973) *(out of print)*
15. *R. Larsen,* Functional Analysis: An Introduction (1973) *(out of print)*
16. *K. Yano and S. Ishihara,* Tangent and Cotangent Bundles: Differential Geometry (1973) *(out of print)*
17. *C. Procesi,* Rings with Polynomial Identities (1973)
18. *R. Hermann,* Geometry, Physics, and Systems (1973)
19. *N. R. Wallach,* Harmonic Analysis on Homogeneous Spaces (1973) *(out of print)*
20. *J. Dieudonné,* Introduction to the Theory of Formal Groups (1973)
21. *I. Vaisman,* Cohomology and Differential Forms (1973)
22. *B.-Y. Chen,* Geometry of Submanifolds (1973)
23. *M. Marcus,* Finite Dimensional Multilinear Algebra (in two parts) (1973, 1975)
24. *R. Larsen,* Banach Algebras: An Introduction (1973)
25. *R. O. Kujala and A. L. Vitter (eds.),* Value Distribution Theory: Part A; Part B: Deficit and Bezout Estimates by Wilhelm Stoll (1973)
26. *K. B. Stolarsky,* Algebraic Numbers and Diophantine Approximation (1974)
27. *A. R. Magid,* The Separable Galois Theory of Commutative Rings (1974)
28. *B. R. McDonald,* Finite Rings with Identity (1974)
29. *J. Satake,* Linear Algebra (S. Koh, T. A. Akiba, and S. Ihara, translators) (1975)

30. *J. S. Golan,* Localization of Noncommutative Rings (1975)
31. *G. Klambauer,* Mathematical Analysis (1975)
32. *M. K. Agoston,* Algebraic Topology: A First Course (1976)
33. *K. R. Goodearl,* Ring Theory: Nonsingular Rings and Modules (1976)
34. *L. E. Mansfield,* Linear Algebra with Geometric Applications: Selected Topics (1976)
35. *N. J. Pullman,* Matrix Theory and Its Applications (1976)
36. *B. R. McDonald,* Geometric Algebra Over Local Rings (1976)
37. *C. W. Groetsch,* Generalized Inverses of Linear Operators: Representation and Approximation (1977)
38. *J. E. Kuczkowski and J. L. Gersting,* Abstract Algebra: A First Look (1977)
39. *C. O. Christenson and W. L. Voxman,* Aspects of Topology (1977)
40. *M. Nagata,* Field Theory (1977)
41. *R. L. Long,* Algebraic Number Theory (1977)
42. *W. F. Pfeffer,* Integrals and Measures (1977)
43. *R. L. Wheeden and A. Zygmund,* Measure and Integral: An Introduction to Real Analysis (1977)
44. *J. H. Curtiss,* Introduction to Functions of a Complex Variable (1978)
45. *K. Hrbacek and T. Jech,* Introduction to Set Theory (1978)
46. *W. S. Massey,* Homology and Cohomology Theory (1978)
47. *M. Marcus,* Introduction to Modern Algebra (1978)
48. *E. C. Young,* Vector and Tensor Analysis (1978)
49. *S. B. Nadler, Jr.,* Hyperspaces of Sets (1978)
50. *S. K. Segal,* Topics in Group Rings (1978)
51. *A. C. M. van Rooij,* Non-Archimedean Functional Analysis (1978)
54. *L. Corwin and R. Szczarba,* Calculus in Vector Spaces (1979)
53. *C. Sadosky,* Interpolation of Operators and Singular Integrals: An Introduction to Harmonic Analysis (1979)
54. *J. Cronin,* Differential Equations: Introduction and Quantitative Theory (1980)
55. *C. W. Groetsch,* Elements of Applicable Functional Analysis (1980)
56. *I. Vaisman,* Foundations of Three-Dimensional Euclidean Geometry (1980)
57. *H. I. Freedman,* Deterministic Mathematical Models in Population Ecology (1980)
58. *S. B. Chae,* Lebesgue Integration (1980)
59. *C. S. Rees, S. M. Shah, and C. V. Stanojević,* Theory and Applications of Fourier Analysis (1981)
60. *L. Nachbin,* Introduction to Functional Analysis: Banach Spaces and Differential Calculus (R. M. Aron, translator) (1981)
61. *G. Orzech and M. Orzech,* Plane Algebraic Curves: An Introduction Via Valuations (1981)
62. *R. Johnsonbaugh and W. E. Pfaffenberger,* Foundations of Mathematical Analysis (1981)
63. *W. L. Voxman and R. H. Goetschel,* Advanced Calculus: An Introduction to Modern Analysis (1981)
64. *L. J. Corwin and R. H. Szcarba,* Multivariable Calculus (1982)
65. *V. I. Istrăţescu,* Introduction to Linear Operator Theory (1981)
66. *R. D. Järvinen,* Finite and Infinite Dimensional Linear Spaces: A Comparative Study in Algebraic and Analytic Settings (1981)

67. *J. K. Beem and P. E. Ehrlich,* Global Lorentzian Geometry (1981)
68. *D. L. Armacost,* The Structure of Locally Compact Abelian Groups (1981)
69. *J. W. Brewer and M. K. Smith, eds.,* Emmy Noether: A Tribute to Her Life and Work (1981)
70. *K. H. Kim,* Boolean Matrix Theory and Applications (1982)
71. *T. W. Wieting,* The Mathematical Theory of Chromatic Plane Ornaments (1982)
72. *D. B. Gauld,* Differential Topology: An Introduction (1982)
73. *R. L. Faber,* Foundations of Euclidean and Non-Euclidean Geometry (1983)
74. *M. Carmeli,* Statistical Theory and Random Matrices (1983)
75. *J. H. Carruth, J. A. Hildebrant, and R. J. Koch,* The Theory of Topological Semigroups (1983)
76. *R. L. Faber,* Differential Geometry and Relativity Theory: An Introduction (1983)
77. *S. Barnett,* Polynomials and Linear Control Systems (1983)
78. *G. Karpilovsky,* Commutative Group Algebras (1983)
79. *F. Van Oystaeyen and A. Verschoren,* Relative Invariants of Rings: The Commutative Theory (1983)
80. *I. Vaisman,* A First Course in Differential Geometry (1984)
81. *G. W. Swan,* Applications of Optimal Control Theory in Biomedicine (1984)
82. *T. Petrie and J. D. Randall,* Transformation Groups on Manifolds (1984)
83. *K. Goebel and S. Reich,* Uniform Convexity, Hyperbolic Geometry, and Nonexpansive Mappings (1984)
84. *T. Albu and C. Năstăsescu,* Relative Finiteness in Module Theory (1984)
85. *K. Hrbacek and T. Jech,* Introduction to Set Theory, Second Edition, Revised and Expanded (1984)
86. *F. Van Oystaeyen and A. Verschoren,* Relative Invariants of Rings: The Noncommutative Theory (1984)
87. *B. R. McDonald,* Linear Algebra Over Commutative Rings (1984)
88. *M. Namba,* Geometry of Projective Algebraic Curves (1984)
89. *G. F. Webb,* Theory of Nonlinear Age-Dependent Population Dynamics (1985)
90. *M. R. Bremner, R. V. Moody, and J. Patera,* Tables of Dominant Weight Multiplicities for Representations of Simple Lie Algebras (1985)
91. *A. E. Fekete,* Real Linear Algebra (1985)
92. *S. B. Chae,* Holomorphy and Calculus in Normed Spaces (1985)
93. *A. J. Jerri,* Introduction to Integral Equations with Applications (1985)
94. *G. Karpilovsky,* Projective Representations of Finite Groups (1985)
95. *L. Narici and E. Beckenstein,* Topological Vector Spaces (1985)
96. *J. Weeks,* The Shape of Space: How to Visualize Surfaces and Three-Dimensional Manifolds (1985)
97. *P. R. Gribik and K. O. Kortanek,* Extremal Methods of Operations Research (1985)
98. *J.-A. Chao and W. A. Woyczynski, eds.,* Probability Theory and Harmonic Analysis (1986)
99. *G. D. Crown, M. H. Fenrick, and R. J. Valenza,* Abstract Algebra (1986)
100. *J. H. Carruth, J. A. Hildebrant, and R. J. Koch,* The Theory of Topological Semigroups, Volume 2 (1986)

101. *R. S. Doran and V. A. Belfi,* Characterizations of C*-Algebras: The Gelfand-Naimark Theorems (1986)
102. *M. W. Jeter,* Mathematical Programming: An Introduction to Optimization (1986)
103. *M. Altman,* A Unified Theory of Nonlinear Operator and Evolution Equations with Applications: A New Approach to Nonlinear Partial Differential Equations (1986)
104. *A. Verschoren,* Relative Invariants of Sheaves (1987)
105. *R. A. Usmani,* Applied Linear Algebra (1987)
106. *P. Blass and J. Lang,* Zariski Surfaces and Differential Equations in Characteristic $p > 0$ (1987)
107. *J. A. Reneke, R. E. Fennell, and R. B. Minton.* Structured Hereditary Systems (1987)
108. *H. Busemann and B. B. Phadke,* Spaces with Distinguished Geodesics (1987)
109. *R. Harte,* Invertibility and Singularity for Bounded Linear Operators (1988).
110. *G. S. Ladde, V. Lakshmikantham, and B. G. Zhang,* Oscillation Theory of Differential Equations with Deviating Arguments (1987)
111. *L. Dudkin, I. Rabinovich, and I. Vakhutinsky,* Iterative Aggregation Theory: Mathematical Methods of Coordinating Detailed and Aggregate Problems in Large Control Systems (1987)
112. *T. Okubo,* Differential Geometry (1987)
113. *D. L. Stancl and M. L. Stancl,* Real Analysis with Point-Set Topology (1987)
114. *T. C. Gard,* Introduction to Stochastic Differential Equations (1988)
115. *S. S. Abhyankar,* Enumerative Combinatorics of Young Tableaux (1988)
116. *H. Strade and R. Farnsteiner,* Modular Lie Algebras and Their Representations (1988)
117. *J. A. Huckaba,* Commutative Rings with Zero Divisors (1988)
118. *W. D. Wallis,* Combinatorial Designs (1988)
119. *W. Więsław,* Topological Fields (1988)
120. *G. Karpilovsky,* Field Theory: Classical Foundations and Multiplicative Groups (1988)
121. *S. Caenepeel and F. Van Oystaeyen,* Brauer Groups and the Cohomology of Graded Rings (1989)
122. *W. Kozlowski,* Modular Function Spaces (1988)
123. *E. Lowen-Colebunders,* Function Classes of Cauchy Continuous Maps (1989)

Other Volumes in Preparation

Function Classes of Cauchy Continuous Maps

EVA LOWEN-COLEBUNDERS
Free University of Brussels
Brussels, Belgium

MARCEL DEKKER, INC. New York and Basel

LIBRARY OF CONGRESS
Library of Congress Cataloging-in-Publication Data

Lowen-Colebunders, Eva
Function classes of Cauchy continuous maps / Eva Lowen
-Colebunders.
p. cm. -- (Monographs and textbooks in pure and applied
mathematics ; 123)
Bibliography: p.
Includes index.
ISBN 0-8247-7992-4
1. Functions, Continuous. 2. Mappings (Mathematics) I. Title.
II. Title: Cauchy continuous maps. III. Series: Monographs and
textbooks in pure and applied mathematics ; v. 123.
QA320.L65 1989
515.7--dc19
88-22700
CIP

MARCEL DEKKER, INC.
270 Madison Avenue, New York, New York 10016

Current printing (last digit):
10 9 8 7 6 5 4 3 2 1

PRINTED IN THE UNITED STATES OF AMERICA

TO THE MEMORY OF MY BELOVED FATHER,
GASTON COLEBUNDERS,
WHOSE IDEALISM AND DEVOTION TO HIS WORK
WILL ALWAYS BE MY GUIDING EXAMPLE

Preface

A real-valued function on a Hausdorff uniform space $(X,\underline{U})$ is said to be Cauchy continuous if it maps Cauchy filters on $(X,\underline{U})$ to Cauchy filters on $\mathbb{R}$. It is well known that this condition of Cauchy continuity is necessary and sufficient for the function to have a continuous extension to the uniform completion $(\hat{X},\hat{\underline{U}})$. Consequently, if we are interested in continuous extensions, we should start with Cauchy continuous functions rather than uniformly continuous functions. Moreover, many functions occurring in analysis are Cauchy continuous but not uniformly continuous. For example, polynomials of degree ≥ 2 are Cauchy continuous on every subset of the real line, whereas, for instance, they are not uniformly continuous on half lines. On any pre-Hilbert space X the inner product $\langle\ ,\ \rangle : X \times X \to \mathbb{K}$ is Cauchy continuous, but it is uniformly continuous only when X is trivial. Many other examples of Cauchy continuous functions that are not uniformly continuous appear in analysis, for instance, in distribution theory (Ref. 110). In the framework of uniform spaces described above, Snipes [108] studied properties of Cauchy continuous maps (called Cauchy regular maps there) and illustrated their advantages over uniformly continuous maps.

The way we deal with Cauchy continuous functions is more global in the sense that we shall study properties of the class $\Gamma(X)$ of real-valued Cauchy continuous maps on X, rather than the properties of the functions in $\Gamma(X)$. Many examples of such classes $\Gamma(X)$ are very familiar. Every realcompactification of a completely regular space X can be obtained by completing a suitable compatible uniformity, and thus it gives rise to a function class $\Gamma(X)$ of Cauchy continuous maps on X. In particular, the Hewitt realcompactification υX gives rise to C(X), but for all other realcompactifications we obtain for $\Gamma(X)$ a proper subclass of C(X).

In the study of properties of the class $\Gamma(X)$ another advantage of Cauchy continuous maps will become clear. The class $\Gamma(X)$ induced by the uniform space $(X,\underline{U})$ is an algebra, whereas the class U(X) of all real-valued uniformly continuous functions in general is not. We shall prove that $\Gamma(X)$ even has much stronger "composition" properties, and that, to a certain extent, it behaves as well as classes of continuous maps. For instance, it will be possible to develop a nice structure space theory on the class $\Gamma(X)$. On U(X), which is not even an algebra, this would of course be impossible.

It is clear that, for the class $\Gamma(X)$ induced by a uniform space $(X,\underline{U})$ some of the information contained in the uniform structure is superfluous. The class $\Gamma(X)$ is completely determined by the $\underline{U}$-Cauchy filters. Also, $\Gamma(X)$ is completely determined by the topology of the completion $(\hat{X},\hat{\underline{U}})$. With this idea in mind let us now consider the following more general but still very common situation. If X is a Hausdorff topological space and Y is any topological Hausdorff extension of X, then Y induces on X the class of all real-valued functions that are continuously extendible to all of Y. For example, every H-closed extension [63], every T_3 extension [62], every e-compact extension [53], and every cocompact extension [1], although not necessarily constructible in the uniform setting, gives rise to such a class of continuously extendible functions. To study this function class, we need an explicit structure on X that contains enough information on Y to express continuous extendibility. The nearness subspace structure on X induced by Y can be used for this purpose. This was shown by Herrlich in [56]. His theorem implies that a real-valued function on X is continuously extendible to Y if and only if the function maps every near collection on X to a near collection on the real line.

In this work we generalize that idea yet one step further. Let X be a Hausdorff convergence space and Y any Hausdorff extension of X. Again the extension Y induces on X the class of all real-valued functions that are continuously extendible to all of Y. For example, every compact extension [73], every T-regular closed extension [77], and every cocomplete extension [81] gives rise to such a function class. It is our purpose to study these function classes on X arising from general Hausdorff convergence extensions. In view of the fact that a convergence space is not necessarily a nearness space, a nearness subspace structure is not available in this general case. A natural subspace structure to take in the setting of convergence spaces is the Cauchy subspace structure. In Chapter 1 we show that the Cauchy subspace structure induced by Y on X is indeed rich enough to express continuous extendibility. We prove that a real-valued function on X is continuously extendible to the strict extension Y if and only if it is Cauchy continuous, meaning that it maps Cauchy filters (of the induced Cauchy space) to Cauchy filters on the real line.

With regard to the internal description of continuous extendibility discussed above the following question now arises naturally: What is the relation between the nearness subspace structure and the Cauchy subspace structure of a topological extension Y, and what is the relation with the collection of Cauchy filters induced by the uniformity when Y is a uniform extension? We answer this question in Chapter 2. To do this we look for a convenient category that contains as subcategories the categories Near of nearness spaces, Unif of uniform spaces, and Chy of Cauchy spaces. Most of the material presented in Chapter 2 was obtained by the author in collaboration with H. L. Bentley and H. Herrlich, and it is part of Ref. 14.

While the first two chapters contain the motivation for our work, namely, the significance of Cauchy continuous maps as morphisms in the category of Cauchy spaces on one hand, and as continuously extendible functions on the other hand, Chapter 3 deals with the properties of function classes of Cauchy continuous maps. We start with an arbitrary Cauchy space X and we let $\Gamma(X)$ be the collection of Cauchy continuous real-valued functions on X. We show that $\Gamma(X)$ is a unitary, uniformly complete algebra and lattice. Moreover, $\Gamma(X)$ is composition closed in the sense of Császár [31, 32] and is closed under bounded inversion in the sense of Ref. 61. However, in general $\Gamma(X)$ is not inversion closed [61].

This fact already illustrates a point of divergence when we compare $\Gamma(X)$ to collections C(X) of continuous maps. This comparison is the subject of Chapter 4. There we characterize the classes C(X) among the classes $\Gamma(X)$. Strongly composition closedness or completeness for the Cauchy structure of continuous convergence are characteristic of C(X). Other characterizations of the classes C(X) among the classes $\Gamma(X)$ are obtained as combinations of the following properties: inversion closedness, sequential closedness, and properties of the separation of zero sets. Further, we investigate the effect of completeness and realcompactness of the given Cauchy space X and its compatible convergence space on the equality of $\Gamma(X)$ and $C(X)$. In the same chapter we then develop structure space theory for $\Gamma(X)$. We show that to a certain extent the structure spaces of $\Gamma(X)$ are similar to the spaces βX and υX built from $C(X)$. The divergence of the two theories is created mainly by the lack of parallelism between ideals in $\Gamma(X)$ and ultrafilters on the zero sets of $\Gamma(X)$. This, in turn, arises from the difference between $\Gamma(X)$ and $C(X)$ which we stressed earlier, namely, that $\Gamma(X)$ is in general not inversion closed.

In Chapter 5 we give necessary and sufficient conditions for a function class Φ on a set X to be a collection $\Gamma(X)$ for a suitable Cauchy structure on X. We show that composition closedness is such a characterizing condition for Φ. We introduce a new completeness notion for a function class Φ by means of a natural Cauchy structure on Φ, which is stronger than uniform completeness and which is the clue for a second characterization. We prove that a point-separating function class is a collection of the type $\Gamma(X)$ if and only if it is a complete function algebra containing the constants. By means of a combination of our first and second characterizations of the function classes $\Gamma(X)$ we answer the following problem, posed by A Császár in Ref. 32: Give a characterization of composition closed function classes by means of finite composition properties and a suitable completeness notion.

In Chapter 6 we start with a function class Φ on either a topological space or a convergence space. From the results of Chapter 5, it follows that composition closedness of Φ guarantees the existence of a Cauchy structure $\underline{C}$ on the set X such that $\Gamma(X) = \Phi$. Of course, when a structure is already given on X, it is natural to ask that the Cauchy structure $\underline{C}$ be compatible with it. In this new context, composition closedness of Φ is no longer sufficient. The problem of finding necessary and sufficient conditions on Φ such that $\Phi = \Gamma(X)$ for some compatible Cauchy structure $\underline{C}$ will be specified somewhat further by asking that the Cauchy structure $\underline{C}$ satisfy some preconceived conditions, such as having a Hausdorff topological completion or having a c-embedded completion. In this form our solution can be used to solve some cases of the following general problem posed by Császár [33]: Let X be a topological space and Φ a class of real-valued functions on X. Given a class T of topological spaces, find necessary and/or sufficient conditions that there exist an extension Y of X that belongs to T and such that Φ consists of those functions that have a continuous extension to Y. Császár has given solutions for the class of all completely regular spaces [32], for the class of all topological spaces, for the class of all T_1 spaces, and for the class of all Hausdorff spaces [33]. In Ref. 87 we obtained solutions for the class of all relatively ω-regular reciprocal topological spaces. In Chapter 6, applying the Cauchy space formulation of the problem, we obtain solutions for several other classes of spaces. Finally, our conditions will be compared with the conditions of Császár.

The material in Chapters 5 and 6 illustrates the importance of Cauchy spaces in the theory of extensions and in the theory of function spaces. These applications of Cauchy spaces lead to simple solutions of various topological problems. These and other applications of Cauchy spaces in various subjects, such as in completion theory of uniform convergence spaces and convergence vectorspaces [46, 78, 95, 100], in the theory of C* algebras [89], and lattice theory [3, 4], confirm our conviction that interest in the theory of Cauchy spaces will become more widespread. It is our hope that this monograph is a contribution in this direction.

Chapters 2–6 are based on the author's Habilitation thesis defended at the Vrije Universiteit Brussel in January 1986 [88]. I wish to thank the members of the jury for their cooperation. In particular, I am grateful to Prof. P. Wuyts (V.U.B. Brussels and RUCA Antwerp) for the support I have been able to count on throughout the years. My special thanks also go to two other members of the jury, Prof. H. L. Bentley (University of Toledo, Ohio) and Prof. H. Herrlich (Universität Bremen) for introducing me to merotopic theory. Their questions gave me new insight into Cauchy spaces, and our collaboration, resulting in Chapter 2, gave me much satisfaction in my mathematics the past three years. I also thank Prof. N. De Grande-De Kimpe (V.U.B. Brussels) for her questions that led to Example 3.4.11. My sincere thanks go also to Prof. D. Kent (Washington State University) for the many friendly letters I received during the past ten years. Exchanging ideas on Cauchy spaces was an invaluable stimulation for my work. I am indebted to Prof. A. Császár (Eötvös Lorand University, Hungary) for the correspondence on problems which led to the development of Chapter 6. The Mathematische Gesellschaft of the DDR and the Mathematical Institute of the Czechoslovak Academy of Sciences are acknowledged for stimulating the development of convergence theory. I have benefited much from their nicely organized conferences.

Eva Lowen-Colebunders

Contents

Function Classes of Cauchy Continuous Maps

1

Extensions and Continuously Extendible Functions

In this chapter we shall develop the material on convergence spaces and Cauchy spaces. We shall restrict ourselves to notions and results needed in the rest of the work. Cauchy continuous maps are introduced here as the morphisms in the category of Cauchy spaces.

Of particular interest for our work is the relation between the categories Conv of all convergence spaces and Chy of all Cauchy spaces. In particular, we develop the result that the category Conv_H of Hausdorff convergence spaces is a bicoreflective and bireflective full subcategory of the category of T_1 Cauchy spaces. Moreover, in this relation Hausdorff convergence extensions correspond to Cauchy completions. That is why we develop part of the theory of Cauchy completions in Section 1.4. The impact of this theory on continuously extendible functions lies in the extension theorem (Theorem 1.4.12) of Section 1.4, from which it is clear that for strict Hausdorff extensions, continuous extendibility is essentially Cauchy continuity.

1.1 SET THEORETIC CONVENTIONS

In general we shall adopt the terminology of N. Bourbaki [24]. Further, if X is a set, we use the following notation.

1.1.1 Notation

$\underline{P}(X)$ is the set of all subsets of X.
$\underline{P}^2(X)$ is the set of all subsets of $\underline{P}(X)$.

If $\underline{A} \subset \underline{P}(X)$, then

sec $\underline{A}$ is the set of all subsets of X intersecting all members of $\underline{A}$

and

$\text{stack}_X\underline{A}$ is the set of all subsets of X containing some member of $\underline{A}$

If $x \in X$, then

$$\dot{x} = \text{stack}_X \{\{x\}\}$$

If $\underline{A} \subset \underline{P}(X)$, $\underline{B} \subset \underline{P}(X)$, then

$\underline{A} < \underline{B}$ ($\underline{A}$ corefines $\underline{B}$) means every member of $\underline{A}$ contains some member of $\underline{B}$

If $f : X \to Y$ is a function, $\underline{A} \subset \underline{P}(X)$ and $A \in \underline{A}$, then

$$f(A) = \{f(x) \mid x \in A\}$$
$$f(\underline{A}) = \{f(A) \mid A \in \underline{A}\}$$

If $\underline{B} \subset Y^X$ and $A \subset X$, then

$$\underline{B}(A) = \{f(x) \mid f \in \underline{B},\ x \in A\}$$

If $\Theta \subset \underline{P}(Y^X)$ and $\underline{A} \subset \underline{P}(X)$, then

$$\Theta(\underline{A}) = \{\underline{B}(A) \mid \underline{B} \in \Theta,\ A \in \underline{A}\}$$

If $\Theta = \{\{ f \}\}$, then we write

$$\Theta(\underline{A}) = f(\underline{A})$$

1.1.2 Notation

A filter on X always is a proper filter.
F(X) is the collection of all filters on X.
U(X) is the collection of all ultrafilters on X.
If $(x_n)_{n \in N}$ is a sequence, then $< x_n >$ is stack $\{\{x_m \mid m \geq n\} \mid n \in N\}$. It is the Frĕchet filter of the sequence $(x_n)_{n \in N}$.

1.1.3 Proposition On F(X) the relation $\underline{F} < \underline{G}$ coincides with the inclusion $\underline{F} \subset \underline{G}$. When F(X) is ordered by inclusion it is a complete inf-lattice and a conditionally complete sup-lattice [23], that is, we have the following operations on F(X):

If $\{\underline{F}_i \mid i \in I\}$ is a collection of filters, then $\{\cup_{i \in I} F_i \mid F_i \in \underline{F}_i\}$ is a filter, denoted by $\cap_{i \in I} \underline{F}_i$ or $\inf_{i \in I} \underline{F}_i$.

If $\cap_{i \in I} F_i \neq \emptyset$ for every finite subset $J \subset I$ and $F_i \in \underline{F}_i$, then $\{\cap_{i \in J} F_i \mid F_i \in \underline{F}_i,\ J \subset I \text{ finite}\}$ again is a filter, denoted by $\sup_{i \in I} \underline{F}_i$. In this case we will write $\sup_{i \in I} \underline{F}_i$ exists.

If $\cap_{i \in J} F_i = \emptyset$ for some finite subset J of I and $F_i \in \underline{F}_i$, then we say that $\sup_{i \in I} \underline{F}_i$ does not exist.

1.2 CONVERGENCE SPACES, UNIFORM CONVERGENCE SPACES

The theory of convergence structures was founded by Choquet [28], Kowalsky [79], Fisher [35], and Kent [67]. Most of the notions below are taken from these papers. Some others are taken from more advanced papers such as [21] and [103].

1.2.1 Definition In our context a convergence structure is a map

$$q : X \to \underline{P}^{F(X)} : x \to q(x)$$

that assigns to every point of X a collection of filters on X. $\underline{F} \in q(x)$ will also be written "$\underline{F}$ q-converges to x" or "x is a q-limit point of $\underline{F}$," or, if no confusion can result, simply "$\underline{F}$ converges to x" or "x is a limit point of $\underline{F}$." The set of all limit points of a given filter $\underline{F}$ will be denoted by lim $\underline{F}$.

The map q is subject to the following axioms.

1. x converges to x.
2. If $\underline{F}$ converges to x and $\underline{G}$ is a filter, $\underline{G} > \underline{F}$, then $\underline{G}$ converges to x.
3. If $\underline{F}$ and $\underline{G}$ both converge to x, then $\underline{F} \cap \underline{G}$ converges to x.

The couple (X,q) of the set X endowed with the convergence structure q is called a convergence space or simply a space. Sometimes we simply write X to denote both the set and the convergence structure on it.

Note that these structures are called "limitierung" in [35] and "pseudotopology" in [45]. Next we define the morphisms between two convergence spaces.

1.2.2 Definition A function $f : (X,q) \to (Y,p)$ between two convergence spaces is <u>continuous</u> if we have

If $\underline{F}$ q-converges to x, then $\text{stack}_Y f(\underline{F})$ p-converges to f(x)

Let Conv be the concrete category of convergence spaces and continuous maps. Conv is a topological category in the sense of [57]. In particular, Conv has initial and final structures, so it is complete and cocomplete. It follows that the convergence structures on a set form a complete lattice and that Conv has subobjects, products, quotients, and coproducts.

Conv is cartesian closed, and the "natural" convergence structure on C(X,Y), the set of all continuous functions from X to Y, is <u>continuous convergence</u>, defined as follows.

A filter Θ on C(X,Y) converges to f if and only if $\text{stack}_Y \Theta(\underline{F})$ converges to f(x) whenever $\underline{F}$ converges to x

On C(X), the set of real-valued continuous functions on X, continuous convergence of sequences of functions was introduced in [51]. The filter description given above and properties of this (nontopological) convergence can be found in [29]. An extensive study of the space C(X) endowed with continuous convergence has been made by the school of Binz [21]. Cartesian closedness of Conv was shown in [25], [29], and [96].

1.2.3 Definition Pstop is the full subcategory of Conv whose objects are the <u>pseudotopological convergence spaces</u>, that is, convergence spaces satisfying the following stronger axiom.

3.' If every ultrafilter finer than $\underline{F}$ converges to x then $\underline{F}$ converges to x.

Pretop is the full subcategory of Conv whose objects are the pretopological convergence spaces, that is, convergence spaces satisfying the following stronger axiom.

3." $\cap \{ \underline{F} \in F(X) \mid \underline{F}$ converges to $x\}$ converges to x.

Pretopological spaces are also known as closure spaces [27] or principal convergence spaces [35].

Clearly every topological space is a (pretopological) convergence space and Top, the category of topological spaces, is isomorphic to a full subcategory of Pretop. However, many examples are known of nontopological convergence spaces: order convergence (when axiom (3) is slightly weakened) [68,113], continuous convergence [21,22,29], closed convergence [28,42,43,82, 92], convergence of Mikusinsky operators [91,111], local uniform convergence [97,98], and Mackey convergence [41] are only a few of the many well known examples.

The relationship between the subcategories of Conv is summarized in the following diagram:

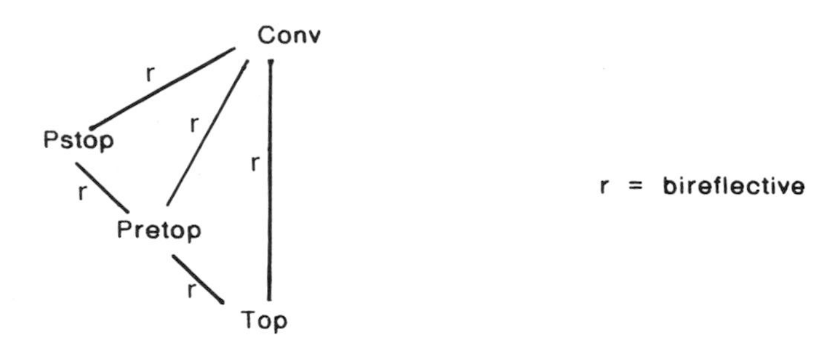

Details on the construction of the bireflections can be found in [35] and [67]. We give a short formulation below.

1.2.4 Proposition If (X,q) is a convergence space, then the Pstop reflection of (X,q) is $1_X : (X,q) \to (X,\tilde{q})$ where $(X,\tilde{q})$ is the convergence space defined by:

$\underline{F}$ $\tilde{q}$-converges to x if and only if every ultrafilter finer than $\underline{F}$ q-converges to x

The definition of the reflector Conv-Pretop is based on the following notions.

1.2.5 Definition If (X,q) is a convergence space and $\underline{F} \in F(X)$, we put $\alpha_q\underline{F} = \{ x \in X \mid \exists\, \underline{G} \in F(X),\ \underline{G} > \underline{F},\ \underline{G}$ q-converges to x$\}$ and we call it the set of <u>adherence points</u> of $\underline{F}$.

In the particular case when $\underline{F}$ = stack{ A } for some subset A of X, then we write

$cl_q A$ instead of α_q stack{ A }

This set is called the <u>closure</u> of A.

The <u>interior</u> of a set A is the set $int_q A$ defined by

$$int_q A = X \setminus cl_q(X \setminus A)$$

The <u>neighborhoodfilter</u> of a point x is the filter $\underline{B}_q(x)$ defined by

$$\underline{B}_q(x) = \cap \{ \underline{F} \mid \underline{F} \text{ q-converges to } x \}$$

It is clear that we also have the following equalities:

$$\underline{B}_q(x) = \{ B \subset X \mid x \notin cl_q(X \setminus B) \} = \{ B \subset X \mid x \in int_q B \}$$

1.2.6 Proposition If (X,q) is a convergence space, then the Pretop reflection of (X,q) is $1_X : (X,q) \to (X,\psi q)$ where $(X,\psi q)$ is the convergence space defined by:

$\underline{F}$ ψq-converges to x if and only if $\underline{F} > \underline{B}_q(x)$

The reflector Conv-Top is based on the following notions.

1.2.7 Definition A subset A of (X,q) is called

q-closed if $cl_q A = A$
q-open if $int_q A = A$

1.2.8 Proposition If (X,q) is a convergence space, then the Top reflection of (X,q) is $1_X : (X,q) \to (X, \lambda q)$ where $(X, \lambda q)$ is the topological space defined by:

$$\lambda q = \{ A \mid A \subset X, \text{ A q-open} \}$$

Other subcategories of Conv are defined by means of symmetry, separation, regularity, or compactness conditions for convergence spaces.

1.2.9 Definition A convergence space (X,q) can have the following additional properties:

T_1 if $\dot{x}$ converges to y implies x = y
Hausdorff if $\underline{F}$ converges to x and y implies x = y
Symmetric if $\dot{x}$ converges to y implies x and y have the same convergent filters
Reciprocal if $\underline{F}$ converges to x and y implies x and y have the same convergent filters

The implications between these properties are summarized in the following diagram:

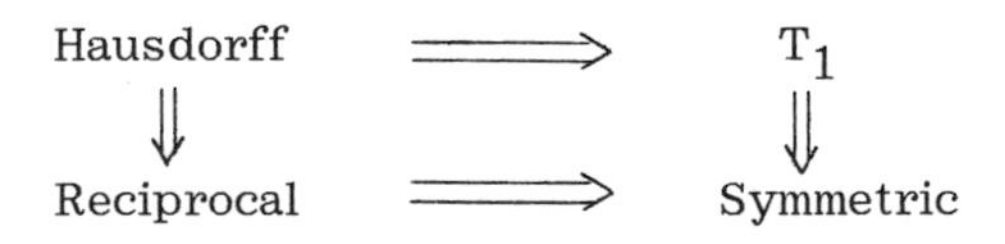

Note that a convergence space is Hausdorff if and only if it is T_1 and reciprocal.

The full subcategories of Conv whose objects are the Hausdorff, symmetric, or reciprocal spaces are denoted by Conv_H, Conv_S, and Conv_R respectively. Further, we let

$$\mathrm{Pstop}_H = \mathrm{Pstop} \cap \mathrm{Conv}_H$$
$$\mathrm{Pretop}_H = \mathrm{Pretop} \cap \mathrm{Conv}_H$$
$$\mathrm{Top}_H = \mathrm{Top} \cap \mathrm{Conv}_H = \mathrm{Haus}$$

The subcategories Pstop_S, Pstop_R, Pretop_S, Pretop_R, Top_S and Top_R are all defined analogously.

To define regularity conditions we first need the following notions and results.

1.2.10 Notation For a filter or a filterbase $\underline{F}$ on (X,q) let

$$\mathrm{cl}_q\underline{F} = \mathrm{stack}\,\{\, \mathrm{cl}_q F \mid F \in \underline{F} \,\}$$

1.2.11 Proposition Let $C(X)$ be the collection of continuous maps from the convergence space (X,q) to the real line with the usual topology and let ωq be the initial topology on X determined by $C(X)$. Then $1_X : (X,q) \to (X,\omega q)$ is the reflection of (X,q) in the subcategory of completely regular topological spaces.

1.2.12 Definition A convergence space (X,q) is

regular if $\underline{F}$ converges to x implies $\mathrm{cl}_q\underline{F}$ converges to x
ω-regular if $\underline{F}$ converges to x implies $\mathrm{cl}_{\omega q}\underline{F}$ converges to x

It was shown in [71] that ω-regular, Hausdorff pseudotopological convergence spaces are exactly the c-embedded convergence spaces of E. Binz [21, 28]. The full subcategory of c-embedded convergence spaces is denoted by $\mathrm{Conv}_{c\text{-emb}}$.

1.2.13 Definition A convergence space (X,q) is

<u>compact</u> if $\alpha_q \underline{F} \neq \emptyset$ for every $\underline{F} \in F(X)$

The full subcategory of compact Hausdorff convergence spaces is denoted by $\text{Conv}_{H.\text{comp}}$.

At this point it is useful to recall that the theory of compact Hausdorff convergence spaces deviates in many respects from the theory of compact Hausdorff topological spaces: for example, compact Hausdorff convergence spaces are not necessarily regular [102].

Uniform convergence spaces were introduced by Cook and Fisher in [30] and were studied in [66, 78, 80], and [100]. Wyler [116] introduced a slightly weaker definition which we will use here. First we need some notation.

If Φ is a filter on $X \times X$, then $\Phi^{-1} = \{F^{-1} \mid F \in \Phi\}$.
If Φ and Ψ are filters on $X \times X$, then
$\Phi \circ \Psi = \text{stack}\,\{F \circ G \mid F \in \Phi, G \in \Psi\}$ whenever
$F \circ G \neq \emptyset$ for every $F \in \Phi$, $G \in \Psi$, that is, whenever
$\Phi \circ \Psi$ exists.

1.2.14 Definition A <u>uniform convergence structure</u> $\underline{G}$ on X is a set of filters on $X \times X$ such that

1. $\dot{x} \times \dot{x} \in \underline{G}$.
2. $\Phi \in \underline{G}$ and $\Psi > \Phi$ implies $\Psi \in \underline{G}$.
3. $\Phi \in \underline{G}$ and $\Psi \in \underline{G}$ implies $\Phi \cap \Psi \in \underline{G}$.
4. $\Phi \in \underline{G}$ implies $\Phi^{-1} \in G$.
5. $\Phi \in \underline{G}$ and $\Psi \in \underline{G}$ implies $\Phi \circ \Psi \in \underline{G}$ (whenever $\Phi \circ \Psi$ exists).

The set X together with the uniform convergence structure $\underline{G}$ is called a <u>uniform convergence space</u>. The morphisms between uniform convergence spaces are the following.

1.2.15 Definition A map $f: (X,\underline{G}) \to (Y,\underline{D})$ between two uniform convergence spaces is uniformly continuous if and only if

$$f \times f(\underline{G}) \subset \underline{D}$$

The concrete category of uniform convergence spaces and uniformly continuous maps is denoted by UCS. It is a topological category, and it is also cartesian closed. This was shown by Wyler in [116]. The relation between UCS and Conv will be explained in the next paragraph.

1.2.16 Definition A filter $\underline{F}$ on $(X,\underline{G})$ is called a Cauchy filter in the uniform convergence space $(X,\underline{G})$ if

$$\underline{F} \times \underline{F} \in \underline{G}$$

These Cauchy filters play an essential role in the completion theory of uniform convergence spaces [100].

1.3 CAUCHY SPACES

Cauchy spaces first appeared in Keller's paper [66]. In that paper the relation between Cauchy spaces, uniform convergence spaces, and convergence spaces was developed. In the completion theory of uniform convergence spaces and convergence vector spaces, Cauchy spaces play an essential role [46, 78, 95, 100, 115]. This fact explains why most work on Cauchy spaces during the last ten years deals mainly with completions [36, 37, 38, 39, 47, 72, 76, 84, 86]. It is precisely the completion theory for Cauchy spaces that is most useful for our study of extensions and continuous extendibility. The basic theory of Cauchy spaces, in the form to be used here, was developed in [66], [72], and [100]. We will also need some notations from some more advanced papers such as [47], [70], [84], [86], and [116].

1.3.1 Definition A *Cauchy structure* on a set X is a subset $\underline{C}$ of F(X) subject to the following axioms:

1. $\dot{x} \in C$.
2. If $\underline{F} \in \underline{C}$ and $\underline{G} > \underline{F}$, then $\underline{G} \in \underline{C}$.
3. If $\underline{F} \in \underline{C}$, $\underline{G} \in \underline{C}$, and sup $\{\underline{F},\underline{G}\}$ exists, then $\underline{F} \cap \underline{G} \in \underline{C}$.

The couple $(X,\underline{C})$ of the set X and the Cauchy structure $\underline{C}$ is called a *Cauchy space*. The members of $\underline{C}$ are the *Cauchy filters*.

We give another way of describing the third Cauchy space axiom. Let $\underline{C} \subset F(X)$ and let ~ be the following relation on $\underline{C}$:

$$\underline{F} \sim \underline{G} \Leftrightarrow \underline{F} \cap \underline{G} \in \underline{C}$$

1.3.2 Proposition $\underline{C}$ is a Cauchy structure on X if and only if $\underline{C}$ satisfies conditions (1), (2), and (3') ~ is an equivalence relation.

If $\underline{F} \in \underline{C}$, then its Cauchy equivalence class is denoted by $\langle \underline{F} \rangle$. Next we define the morphisms between Cauchy spaces.

1.3.3 Definition A function $f : (X,\underline{C}) \to (Y,\underline{D})$ from a Cauchy space $(X,\underline{C})$ to a Cauchy space $(Y,\underline{D})$ is defined to be *Cauchy continuous* if

$$\underline{F} \in \underline{C} \text{ implies } \text{stack}_Y\, f(\underline{F}) \in \underline{D}$$

Let Chy be the concrete category of Cauchy spaces and Cauchy continuous maps. Chy is a topological category. In particular, it has initial and final structures, and consequently, Chy is complete and cocomplete. Therefore the set of all Cauchy structures on a given set is a complete lattice, and Chy has subobjects, products, quotients, and coproducts.

Initial structures in Chy have a very elegant description. We will not give an explicit formulation now but we will postpone it to the second chapter when we are in a position to describe final structures as well. The reason is that final structures in Chy do not have such an elegant description and that the easiest way to obtain them is by applying the bireflection to the final structure formed in a larger category.

Chy is cartesian closed and the "natural" Cauchy structure on the set $\Gamma(X,Y)$ of all Cauchy continuous functions from X to Y is the Cauchy continuous Cauchy structure [13] defined as follows:

> Θ is a Cauchy filter on $\Gamma(X,Y)$ if and only if $\text{stack}_Y\Theta(\underline{F})$ is a Cauchy filter on Y whenever $\underline{F}$ is a Cauchy filter on X

On the class $\Gamma(X)$ of all Cauchy continuous maps from a given Cauchy space $(X,\underline{C})$ to the real line, endowed with the Cauchy structure consisting of all convergent filters, the Cauchy continuous Cauchy structure is denoted by $\hat{\underline{C}}$. Gazik and Kent made an extensive study of $\hat{\underline{C}}$ in [47].

1.3.4 Definition If $(X,\underline{U})$ is a uniform space, the collection $\underline{C}_{\underline{U}}$ of its Cauchy filters is a Cauchy structure. We say that $\underline{C}_{\underline{U}}$ and $\underline{U}$ are compatible. A given Cauchy structure $\underline{C}$ does not necessarily have a compatible uniformity. $\underline{C}$ is *uniformizable* if it does have a compatible uniformity.

Let Chy_U be the full subcategory of Chy whose objects are the uniformizable Cauchy spaces. Further, let Unif be the category of uniform spaces and let Conv_R and UCS be defined as in Section 1.2. The following diagram summarizes the relation between Chy and these other categories. We will describe the embeddings, and some of the reflections and coreflections.

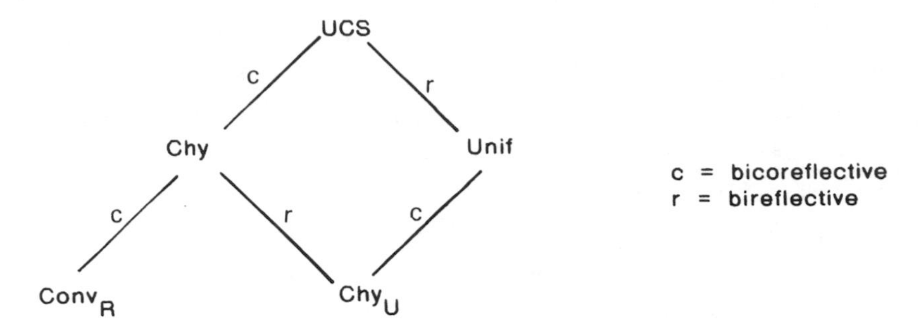

1.3.5 Proposition The category Conv_R is (isomorphic to) a full subcategory of Chy. The embedding is the following:

If (X,q) is a reciprocal convergence space, then

$$\underline{C}^q = \{\underline{F} \in F(X) \mid \underline{F} \text{ q-converges}\}$$

is the corresponding Cauchy structure on X.
Moreover Conv_R is coreflective in Chy.

If $(X,\underline{C})$ is a Cauchy space, then the Conv_R coreflection of $(X,\underline{C})$ is $1_X : (X,q_{\underline{C}}) \to (X,\underline{C})$ where $(X,q_{\underline{C}})$ is the convergence space defined by:

$$\underline{F}\ q_{\underline{C}}\text{-converges to } x \text{ if and only if } \underline{F} \cap \dot{x} \in \underline{C}$$

The Cauchy structure $\underline{C}$ and the convergence structure $q_{\underline{C}}$ are said to be compatible.

1.3.6 Definition A Cauchy space $(X,\underline{C})$ is said to be complete if

$$\underline{F} \in \underline{C} \text{ implies } \underline{F}\ q_{\underline{C}}\text{-converges}$$

So the reciprocal convergence spaces can be identified with the complete Cauchy spaces. Notions such as limit, closure, adherence, openness on a Cauchy space $(X,\underline{C})$ are all defined in the convergence coreflection.

1.3.7 Proposition Chy is (isomorphic to) a full subcategory of UCS. The embedding is the following. If $(X,\underline{C})$ is a Cauchy space then

$$\underline{G}^{\underline{C}} = \{\Phi \in F(X \times X) \mid \exists\, \underline{F}_1, \ldots, \underline{F}_n \in \underline{C},\ \bigcap_{i=1}^{n} \underline{F}_i \times \underline{F}_i < \Phi\}$$

is the corresponding uniform convergence structure on X. Moreover Chy is coreflective in UCS. If $(X,\underline{G})$ is a uniform convergence space then the Chy coreflection of $(X,\underline{G})$ is $1_X : (X,\underline{C}_{\underline{G}}) \to (X,\underline{G})$ where $(X,\underline{C}_{\underline{G}})$ is the Cauchy space defined by

$\underline{F} \in \underline{C}_{\underline{G}}$ if and only if $\underline{F}$ is a $\underline{G}$-Cauchy filter

1.3.8 Proposition 1. Unif is (isomorphic) to a full subcategory of UCS. If $(X,\underline{U})$ is a uniform space then

$$\underline{C}^{\underline{U}} = \{\Phi \in F\,(X \times X) \mid \Phi > \underline{U}\}$$

is the corresponding uniform convergence structure. Moreover Unif is bireflective in UCS.

2. Chy_U, the category of uniformizable Cauchy spaces, is a full subcategory of Chy. It is bireflective in Chy.

3. Chy_U also is isomorphic to a full subcategory of Unif. If $(X,\underline{C})$ is a uniformizable Cauchy space there exists a finest uniformity $\underline{U}^{\underline{C}}$ with $\underline{C}$ as its collection of Cauchy filters. Chy_U is coreflective in Unif and the coreflection is the restriction of the coreflection of UCS in Chy.

Note that the objects in the category $\mathrm{Chy}_U \cap \mathrm{Conv}_H$ are exactly the completely uniformizable Hausdorff topological spaces. We denote this category by Haus_{CU}. Other full subcategories of Chy are defined by imposing some additional properties on the equivalence classes of the Cauchy spaces.

1.3.9 Definition $(X,\underline{C})$ is called

Pseudotopological if $\underline{F} \in \underline{C}$ whenever all ultrafilters finer than $\underline{F}$ belong to the same equivalence class.
Pretopological if every equivalence class has a minimum Cauchy filter.
Topological if every equivalence class has a minimum Cauchy filter with an open base.

Let PsChy be the full subcategory of Chy whose objects are the pseudotopological Cauchy spaces, PreChy the full subcategory whose objects are the pretopological Cauchy spaces, and TopChy the full subcategory whose objects are the topological Cauchy spaces. Other subcategories are defined by means of the following separation, regularity, or compactness conditions.

1.3.10 Definition A Cauchy space $(X,\underline{C})$ is

T_1 if $\dot{x} \cap \dot{y} \in \underline{C}$ implies $x = y$

Note that this T_1 property is essentially a condition on the convergence coreflection $q^{\underline{C}}$. We have:

$$(X,\underline{C}) \text{ is } T_1 \iff (X,q^{\underline{C}}) \text{ is } T_1$$
$$\iff (X,q^{\underline{C}}) \text{ is Hausdorff}$$

T_1Chy is the full subcategory of Chy whose objects are the T_1 Cauchy spaces.

1.3.11 Definition A Cauchy space $(X,\underline{C})$ is

Regular if $\underline{F} \in \underline{C}$ implies $\text{cl}\,\underline{F} \in \underline{C}$

By the usual Cauchy structure on the real line we mean

$\{\underline{F} \in F(R) \mid \underline{F}$ converges in the usual topology$\}$

Unless indicated otherwise, R will be endowed with this structure. If $(X,\underline{C})$ is a Cauchy space, let $\Gamma(X)$ be the collection of all real-valued Cauchy continuous maps. The initial topology on X determined by $\Gamma(X)$ is denoted by $t\Gamma(X)$. This topology is used to define a stronger regularity condition.

1.3.12 Definition $(X,\underline{C})$ is said to be

<u>μ-regular</u> if $\underline{F} \in \underline{C}$ implies $cl_{t\Gamma(X)}\, \underline{F} \in \underline{C}$

<u>Cauchy separated</u> if for every pair $\underline{F},\underline{G}$ of nonequivalent Cauchy filters there is a function $f \in \Gamma(X)$ such that lim stack $f(\underline{F}) \neq$ lim stack $f(\underline{G})$ in R

<u>Totally bounded</u> if $\underline{U} \in U(X)$ implies $\underline{U} \in \underline{C}$

1.4 HAUSDORFF EXTENSIONS

Reciprocal convergence spaces are complete Cauchy spaces, as was established in the previous section. Now we apply this result in using the completion theory of Cauchy spaces to obtain properties of extensions. In particular we shall be dealing with Hausdorff extensions. Recently a nice survey paper on Hausdorff completions of T_1 Cauchy spaces was written by Kent and Richardson [75].

1.4.1 Definition A convergence space (Y,p) is an <u>extension</u> of a convergence space (X,q) if the following conditions are fulfilled:

1. (X,q) is a convergence subspace of (Y,p).
2. X is <u>dense</u> in (Y,p), meaning that $cl_p X = Y$.

(Y,p) is a <u>strict extension</u> of (X,q) if condition (2) is replaced by 2.' X is <u>strictly dense</u> in (Y,p), meaning that, for every y

$\in$ Y and filter $\underline{F}$ on Y, p-converging to y, there exists a filter $\underline{G}$ on Y, containing X and p-converging to y and such that $cl_p\underline{G} < \underline{F}$.

Clearly every strictly dense set is dense and so every strict extension is an extension. Note that topological extensions are always strict.

1.4.2 Definition A <u>completion</u> $((Y,\underline{D}),k)$ of a Cauchy space $(X,\underline{C})$ is a complete Cauchy space $(Y,\underline{D})$ and a Cauchy embedding

$$k: (X,\underline{C}) \to (Y,\underline{D})$$

such that k(X) is dense in Y. We will often identify X with k(X) and consider $(X,\underline{C})$ as a Cauchy subspace of its completion. $(Y,\underline{D})$ is a <u>strict completion</u> of $(X,\underline{C})$ if moreover X is strictly dense in Y.

1.4.3 Proposition If (Y,p) is a (strict) extension of (X,q) and (Y,p) is reciprocal, then its natural Cauchy structure $\underline{C}^p$ induces the Cauchy structure

$$\underline{C} = \{\underline{F} \in F(X) \mid \text{stack}_Y \underline{F} \text{ p-converges}\}$$

on X. Then $(Y,\underline{C}^p)$ is a (strict) completion of $(X,\underline{C})$. Conversely if $(Y,\underline{D})$ is a (strict) completion of a Cauchy space $(X,\underline{C})$, then the coreflection $(Y,q_{\underline{D}})$ is a (strict) extension of $(X,q_{\underline{C}})$. In particular every Hausdorff extension corresponds to a Hausdorff completion of a T_1 Cauchy space and vice versa.

After this section we will restrict ourselves to Hausdorff extensions and to T_1 Cauchy spaces. A quasi-order among the Hausdorff completions of a T_1 Cauchy space $(X,\underline{C})$ is defined in the following way.

1.4.4 Definition If $((Y_1,\underline{D}_1),k_1)$ and $((Y_2,\underline{D}_2),k_2)$ are completions of $(X,\underline{C})$, then $((Y_1,\underline{D}_1),k_1) \leq ((Y_2,\underline{D}_2),k_2)$ if there exists a Cauchy continuous map $\psi\colon (Y_2,\underline{D}_2) \rightarrow (Y,\underline{D}_1)$ such that the following diagram commutes:

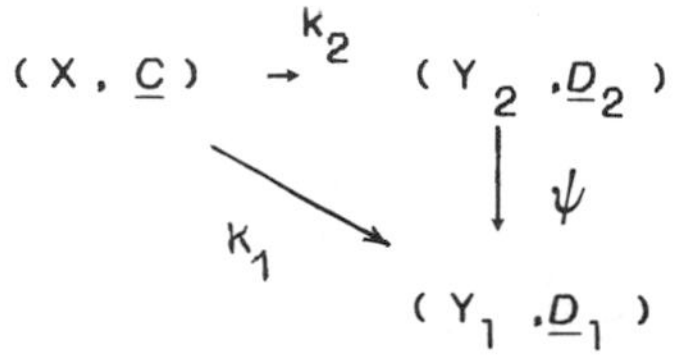

$$\begin{array}{ccc} (X,\underline{C}) & \xrightarrow{k_2} & (Y_2,\underline{D}_2) \\ & \searrow^{k_1} & \downarrow \psi \\ & & (Y_1,\underline{D}_1) \end{array}$$

The completions are equivalent, denoted by $((Y_1,\underline{D}_1),k_1) \simeq ((Y_2,\underline{D}_2),k_2))$ if $((Y_1,\underline{D}_1),k_1) \leq ((Y_2,\underline{D}_2),k_2)$ and $((Y_1,\underline{D}_1),k_1) \leq ((Y_2,\underline{D}_2),k_2)$.

Every T_1 Cauchy space has a huge family of (strict) non-equivalent Hausdorff completions. For our purpose we do not require all explicit constructions of completion methods known. We need only those having nice permanence properties or those playing a particular role with respect to extensions of functions.

Many completion methods use the set of all Cauchy equivalence classes to build the completion. If $(X,\underline{C})$ is a T_1 Cauchy space let

X^* be the set of all Cauchy equivalence classes

The map

$$j\colon X \rightarrow X^*\colon j(x) = \langle \dot{x} \rangle$$

is one-to-one since $(X,\underline{C})$ is T_1.

1.4.5 Definition A Hausdorff completion $((Y,\underline{D}),k)$ of $(X,\underline{C})$ is in standard form if

$$Y = X^* \quad \text{and} \quad k = j$$

In [100] it was shown that every Hausdorff completion is equivalent to one in standard form. One particular class $\underline{R}$ of completions in standard form was constructed by E. Reed [100]. Completions in $\underline{R}$ are all Hausdorff and strict. The construction is described as follows.

1.4.6 Definition Let Λ be the set of all functions $\lambda : X^* \to F(X)$ such that $\lambda j(x) = \dot{x}$ for all $x \in X$, and if $\rho = \langle F \rangle \in X^* \setminus j(X)$, then $\lambda(\rho) < \underline{G}$ for some $\underline{G} \sim \underline{F}$. With each $\lambda \in \Lambda$ and $M \subset X$ let

$$M^{\lambda} = \{\rho \in X^* \mid M \in \lambda(\rho)\}$$

For $\lambda \in \Lambda$ and $\underline{G} \in F(X)$ let

$$\underline{G}^{\lambda} = \text{stack}_{X^*}\{G^{\lambda} \mid G \in \underline{G}\}$$

For each nonempty subset $\Gamma \subset \Lambda$ let $\underline{C}_{\Gamma}$ be the following complete Cauchy structure on X^*:

Ψ converges to ρ if and only if
$\Psi > \underline{G}^{\gamma} \cap \dot{\rho}$ for some $\underline{G} \in \rho$ and $\gamma \in \Gamma$

Let $K_{\Gamma}(X,\underline{C})$ be the completion $((X^*,\underline{C}_{\Gamma}),j)$. Then $\underline{R}$ is the class of all completions $K_{\Gamma}(X,\underline{C})$ for $\Gamma \subset \Lambda$.

This class contains, for instance, the Wyler completion [115] and the Kowalsky completion [79].

1.4.7 Proposition The Wyler completion is obtained by taking $\Gamma = \{\nu\}$ where $\nu(p) = \{X\}$ for $p \notin j(X)$. The resulting completion is denoted by $K_{\nu}(X,\underline{C})$.

In Section 1.3 we described the coreflection

$$1: (X, q_{\underline{C}}) \to (X, \underline{C})$$

of a Cauchy space in to the category of reciprocal convergence spaces. Taking the restriction of this coreflection to T_1Chy we obtain that Conv_H is coreflective in T_1Chy. Every Cauchy continuous map from a T_1 Cauchy space $(X, \underline{C})$ to a complete T_1 Cauchy space has a unique extension to the Wyler completion [37]. This result implies that Conv_H is not only coreflective but also reflective in T_1Chy.

1.4.8 Proposition Conv_H is bireflective in T_1Chy. If $(X, \underline{C})$ is a T_1 Cauchy space, then the bireflection is given by

$$j: (X, \underline{C}) \to K_{\nu}(X, \underline{C})$$

where $K_{\nu}(X, \underline{C})$ is the Wyler completion of $(X, \underline{C})$.

The Wyler completion preserves the pseudotopological, pretopological, or topological property of the given Cauchy space $(X, \underline{C})$. On the other hand it does not necessarily preserve properties such as uniformizability or total boundedness.

1.4.9 Proposition The Kowalsky completion of $(X, \underline{C})$ is the completion $K_{\Sigma}(X, \underline{C})$ where

$$\Sigma = \{\sigma \in \Lambda \mid \sigma(\rho) \in \rho \text{ for all } \rho \in X^*\}$$

This completion preserves a number of important Cauchy properties such as uniformizability and the pseudotopological, pretopological, or topological property of the given Cauchy space $(X, \underline{C})$. It does not necessarily preserve total boundedness.

1.4.10 Definition If $\Gamma = \{\omega\}$ where $\omega \in \Sigma$ and $\omega(\rho)$ is an ultrafilter for every $\rho \in X^*$, then we denote

$K_\omega(X,\underline{C}) = K_\Gamma(X,\underline{C})$

1.4.11 Proposition The completions of type $K_\omega(X,\underline{C})$ preserve total boundedness, uniformizability and the pseudotopological or pretopological property of the given Cauchy space $(X,\underline{C})$. However $K_\omega(X,\underline{C})$ does not necessarily preserve the topological property of $(X,\underline{C})$.

In general, none of the completions in $\underline{R}$ preserve regularity. In fact a regular Cauchy space need not have a (strict) regular Hausdorff completion at all. More details on this topic can be found in [72], [76], and [86]. One very important result in this context is that a strict regular Hausdorff completion, if it exists, is unique up to the usual equivalence. This can be shown directly, as was done in [72], or follows as a corollary from the following extension theorem, the proof of which is a generalization of the extension theorem developed in [100] for completions in $\underline{R}$.

1.4.12 Theorem If $(Y,\underline{D})$ is any strict Hausdorff completion of a T_1 Cauchy space $(X,\underline{C})$, and $(Z,\underline{E})$ is a regular T_1 complete Cauchy space (i.e., a regular T_1 convergence space), then a Cauchy continuous function $f : (X,\underline{C}) \to (Z,\underline{E})$ has a unique (Cauchy) continuous extension $\hat{f} : (Y,\underline{D}) \to (Z,\underline{E})$.

Proof: Let $f : (X,\underline{C}) \to (Z,\underline{E})$ be Caucy continuous. For a point y of Y we use the denseness of X in Y to choose a filter $\underline{F}$ on X such that $\text{stack}_Y\underline{F}$ converges to y in Y, and we define

$$\hat{f}(y) = \lim \text{stack}_Z f(\underline{F})$$

This assignment is independent of the choice of $\underline{F}$. Indeed, if $\underline{G}$ is another filter on X and $\text{stack}_Y\underline{G}$ also converges to y in Y, then $\text{stack}_Y\underline{F} \cap \text{stack}_Y\underline{G}$ converges to y and hence $\underline{F} \cap \underline{G} \in \underline{C}$. Moreover, $\lim \text{stack}_Z f(\underline{F} \cap \underline{G}) = \lim \text{stack}_Z f(\underline{G})$.

The function $\hat{f}$ thus defined is an extension of f, since for $x \in X$ we have

$$\hat{f}(x) = \lim \text{stack}_Z\ f(\dot{x}) = \lim\ (f(x))^{\cdot} = f(x)$$

Finally to prove that f is continuous on Y, let Θ be a filter on Y converging to $y \in Y$. Then by the strict density of X in Y we can choose a filter $\underline{F}$ on X with $\text{stack}_Y\underline{F}$ converging to y and such that $\text{cl}_Y\ \underline{F} < \Theta$. Straightforward verification shows that

$$\text{cl}_Z f(\underline{F}) < \hat{f}(\Theta)$$

The regularity of Z finally ensures that stack $\hat{f}(\Theta)$ converges to $\hat{f}(y)$. Since two continuous functions to a Hausdorff space and coinciding on a dense set of points, are equal, the extension $\hat{f}$ moreover is unique.

Another completion method, "the natural completion," was developed by Gazik and Kent in [47] for the subcategory of all $c^{\wedge}$-embedded T_1 Cauchy spaces.

1.4.13 Definition Let $\Gamma(X,\underline{C})$ be the collection of real-valued Cauchy continuous maps on $(X,\underline{C})$, equipped with the Cauchy continuous structure

$$\hat{\underline{C}} = \{\Theta \in F(\Gamma(X,\underline{C}))\ |\ \text{stack}\ \Theta\ (\underline{F}) \text{ converges whenever } \underline{F} \in \underline{C}\}$$

$\Gamma(\Gamma(X,\underline{C}),\hat{\underline{C}})$ also carries the Cauchy continuous structure $(\hat{\underline{C}})^{\wedge}$. $(X,\underline{C})$ is $c^{\wedge}$-<u>embedded</u> if the evaluation map

$$e: \quad (X,\underline{C}) \to (\Gamma(\Gamma(X,\underline{C}),\hat{\underline{C}}),(\hat{\underline{C}})^{\wedge}) : \ e(x)(f) = f(x)$$

is a Cauchy embedding.

The next results were also shown in [47].

1.4.14 Proposition A Cauchy space is c^-embedded if and only if it is pseudotopological μ-regular, Cauchy separated and T_1.

1.4.15 Proposition For any Cauchy space $(X,\underline{C})$ the space $(\Gamma(X,\underline{C}),\hat{\underline{C}})$ is complete and c^-embedded. If $(X,\underline{C})$ is c^-embedded, then the closure of $e(X)$ in $(\Gamma(\Gamma(X,\underline{C}),\hat{\underline{C}}),(\hat{\underline{C}})\hat{})$ is a c^-embedded completion of $(X,\underline{C})$, called the natural completion of $(X,\underline{C})$ and denoted by $N(X,\underline{C})$.

The natural completion $N(X,\underline{C})$ is always regular and preserves uniformizability and total boundedness. However, it is not necessarily strict and so does not necessarily belong to $\underline{R}$. Also, it does not necessarily preserve the topological property of the given Cauchy space.

The extension theorem [47] can be formulated in the following way.

1.4.16 Proposition The category of c-embedded convergence spaces is (isomorphic to) a full and bireflective subcategory of the category of c^-embedded Cauchy spaces. The embedding is the restriction of the usual one,

$$(X,q) \rightarrow (X,\underline{C}^q)$$

and expresses the correspondence between c-embedded convergence spaces and complete c^-embedded Cauchy spaces. The bireflection is given by

$$e: (X,\underline{C}) \rightarrow N(X,\underline{C})$$

2

The Categories Conv, Chy, Near, and Mer

If Y is a topological completely uniformizable Hausdorff extension of X, then it is well known that the continuously extendible functions are those mapping Cauchy filters of the induced uniformity to convergent filters on the real line. If Y is any topological Hausdorff extension of X, then continuously extendible functions are those mapping near collections of the induced nearness structure to near collections of the real line. This result is a special case of the main theorem in Herrlich's paper [56]. In addition to the previous two well-known extension theorems, we formulated a third one in Theorem 1.4.12. If Y is any strict Hausdorff convergence extension of X, then continuously extendible functions are those mapping Cauchy filters of the induced Cauchy structure to Cauchy filters on the real line. Apparently, in each of the three cases, a different subspace structure is involved.

In this chapter we study the relations between these subspace structures. We first look for a convenient category containing both the categories Near of nearness spaces and Chy of Cauchy spaces. The category Mer of merotopic spaces provides a solution to this problem. We show how Chy, Near, and Unif are embedded in Mer. In particular we are interested in merotopic subspaces of spaces in each of these subcategories. A bonus from embedding Chy and Near in the common supercategory Mer is that in doing so, two theories, convergence and Cauchy theory on one hand, and nearness and merotopic theory on the other, which historically were developed independently from each other, are brought together. It becomes clear that several notions and results, sometimes with different names and terminology, exist in both theories. For instance we will show that separated filter merotopic spaces are exactly the same as the pretopological Cauchy spaces and that separated filter nearness spaces coincide with topological Cauchy spaces.

Most of the material in this chapter was established in collaboration with H. L Bentley and H. Herrlich and is part of the paper "Convergence" [14] by H. L Bentley, H. Herrlich, and the author.

2.1 MEROTOPIC SPACES

Merotopic spaces were first defined by Katětov in [64] by means of micromeric collections. Herrlich [57] showed that merotopic spaces can be defined in any one of three equivalent ways; by means of the micromeric collections, that is, collections that are small, or by means of uniform covers, that is, a generalization of one approach to uniform spaces, or by means of collections that are near, a concept intuitively similar to the one which is involved in proximity spaces.

2.1.1 Definition A <u>merotopic structure</u> on a set X is a subset γ of $\underline{P}^2(X)$ satisfying the following conditions

M1. If $\underline{A} < \underline{B}$ and $\underline{A} \in \gamma$, then $\underline{B} \in \gamma$.
M2. $\dot{x} \in \gamma$ for every $x \in X$.
M3. $\{\emptyset\} \in \gamma$ and $\emptyset \notin \gamma$.
M4. If $\underline{A} \cup \underline{B} \in \gamma$ then $\underline{A} \in \gamma$ or $\underline{B} \in \gamma$.

The members of γ are said to be micromeric collections. γ is a <u>merotopy</u> and (X, γ) is a <u>merotopic space</u>. Explicit mention of the merotopy is sometimes suppressed and we refer to X as the merotopic space.

2.1.2 Definition A function $f: (X,\gamma) \to (Y,\delta)$ between merotopic spaces (X,γ) and (Y,δ) is called <u>uniformly continuous</u> if

$$\underline{A} \in \gamma \text{ implies } f(\underline{A}) \in \delta$$

Mer denotes the concrete category of merotopic spaces and uniformly continuous maps. Mer is a topological category. In particular Mer has initial and final structures and so it is a complete and cocomplete category. Further, the merotopies on a set form a complete lattice, and Mer has subspaces, products, quotients, and coproducts. For an explicit description of these structures we refer to [64].

2.1.3 Definition Let (X, γ) be a merotopic space and let $\underline{F}$ be a filter on X. $\underline{F}$ is said to be a <u>Cauchy filter</u> if $\underline{F}$ is micromeric. (X, γ) is called a <u>filter merotopic space</u> if every micromeric collection is corefined by a Cauchy filter.

The full subcategory of Mer whose objects are all filter merotopic spaces is denoted by Fil. It is clear that a filter merotopic space is completely determined by giving the Cauchy filters. The category Fil was defined by Katětov [64]. It is isomorphic to the category Grill defined by Robertson in [104] and studied by Bentley, Herrlich, and Robertson in [15].

2.1.4 Definition Let C denote the full subcategory of Fil whose objects are those filter merotopic spaces satisfying the following extra condition (C): If $\underline{A}$ and $\underline{B}$ are micromeric in X and if for some point $x \in X$, we have that

$\{A \cup \{x\} \mid A \in \underline{A}\}$ and $\{B \cup \{x\} \mid B \in \underline{B}\}$ are micromeric, then

$\{A \cup B \mid A \in \underline{A}, B \in \underline{B}\}$ is micromeric.

The category C is isomorphic to the subcategory of Grill which Robertson denoted LG in [104]. Recall that a concept of nearness for collections of subsets of a merotopic space can be defined.

2.1.5 Definition If (X, γ) is a merotopic space then a collection $\underline{A}$ of subsets of X is said to be near in (X,γ) provided the collection sec $\underline{A}$ is micromeric.

A merotopic space has an underlying closure operator, which is defined as follows.

2.1.6 Definition For $A \subset X$ we put

$$cl_\gamma A = \{x \in X \mid \{\{x\}, A\} \text{ is near in } (X,\gamma)\}$$

This operator satisfies all conditions of a topological closure operator except for the idempotency.

2.1.7 Definition A merotopic space (X, γ) is said to be a nearness space provided that

$\underline{A}$ is near whenever $\{ cl_X A \mid A \in \underline{A}\}$ is near

An equivalent formulation can be given using covers.

2.1.8 Definition If (X,γ) is a merotopic space, then a collection $\underline{A}$ of subsets of X is called a <u>uniform cover</u> of (X,γ) if the following condition holds.

$$\forall\ \underline{B} \in \gamma \quad \underline{A} \cap \text{stack } \underline{B} \neq \emptyset$$

A merotopic space (X, γ) has an underlying <u>interior operator</u> defined on the subsets of X by

$$\text{int}_\gamma A = \{x \in X \mid \{A, X\backslash\{x\}\} \text{ is a uniform cover}\}$$

This interior operator satisfies all properties of a topological interior operator except for the idempotency. The closure and interior operator are related by

$$x \in \text{int}_\gamma A \Leftrightarrow x \notin cl_\gamma(X\backslash A)$$

An equivalent formulation for nearness spaces is the following.

2.1.9 Proposition A merotopic space is a nearness space provided that

$\{\text{int}_\gamma A \mid A \in \underline{A}\}$ is a uniform cover whenever $\underline{A}$ is a uniform cover

In a nearness space the operators $cl_\gamma A$ and $\text{int}_\gamma A$ both become idempotent.

The full subcategory of Mer whose objects are the nearness spaces is denoted by Near. The category Near was defined and studied by Herrlich [55,57]. From these papers we recall that Top_S, the category of symmetric topological spaces and Unif, the category of uniform spaces, are (isomorphic to) full subcategories of Near.

2.2 SUBCATEGORIES OF MER

We will now establish and explain the indicated relationships between the categories in the following diagram.

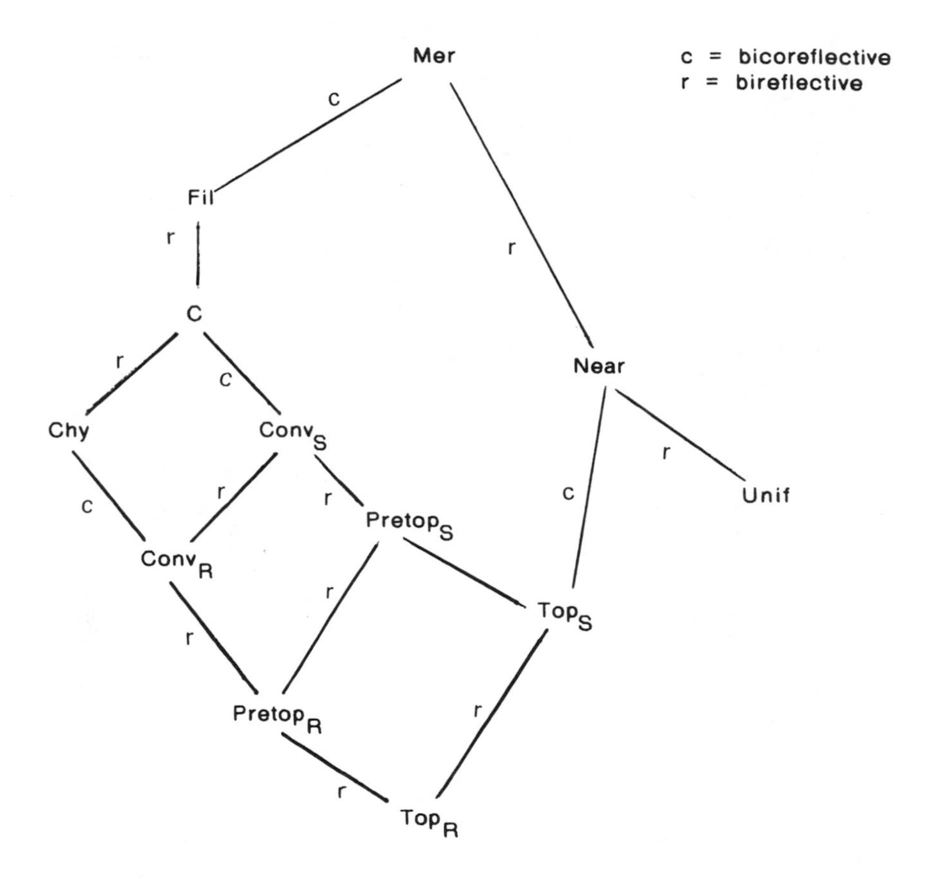

2.2.1 Proposition Fil is bicoreflective in Mer. If (X, γ) is a merotopic space then the Fil coreflection of (X, γ) is the space (X, δ) defined by decreeing that $\underline{A}$ is micromeric on (X, δ), if and only if $\underline{A}$ is corefined by a filter which is Cauchy on (X, γ).

This coreflection corresponds to Katětov's filter modification. In particular, if the filter coreflection is applied to a uniform space $(X, \underline{U})$ then we obtain the uniformizable Cauchy space $(X, \underline{C}_{\underline{U}})$ described in Section 1.3. Being a bicoreflective subcategory of Mer, it follows that final structures in Fil are constructed the same way as in Mer. In fact, Fil also has the same subspaces as Mer. We give an explicit description of the initial and final structures in Fil since they have a particularly elegant formulation and form a basis for constructions of initial and final structures in some subcategories of Fil.

2.2.2 Proposition If $(f_i\colon X \to (X_i, \gamma_i))_{i \in I}$ is a source in Fil, then the initial filter merotopic structure on X is the following:

> A filter $\underline{F}$ on X is a Cauchy filter if and only if $\text{stack}_{X_i} f_i(\underline{F})$ is a Cauchy filter on (X_i, γ_i) for every $i \in I$

If $(f_i\colon (X_i, \gamma_i) \to X)_{i \in I}$ is a sink in Fil then the final filter merotopic structure on X is the following:

> A filter $\underline{F}$ on X is a Cauchy filter if and only if $f_i(\underline{F}_i) < \underline{F}$ for some $i \in I$ and some $\underline{F}_i \in \gamma_i$ or $\dot{x} = \underline{F}$ for some $x \in X$

2.2.3 Proposition C is bireflective in Fil.

The details of the construction of the bireflection can be found in Robertson [104, p. 55]. Being bireflective in Fil, it follows that initial structures in C are constructed the same way as in Fil. In fact, C also has the same coproducts as Fil.

2.2.4 Proposition Chy is (isomorphic to) a full subcategory of C. The embedding is the following. If $(X,\underline{C})$ is a Cauchy space then

$$\gamma_{\underline{C}} = \{\underline{A} \subset \underline{P}^2(X) \mid \underline{F} < \underline{A} \text{ for some } \underline{F} \in \underline{C}\}$$

is the corresponding merotopic structure on X.

The objects of Chy, when embedded in Mer, are exactly the Hausdorff filter merotopic spaces in the sense of Katĕtov [64]. In the next section we shall identify $\underline{C}$ and $\gamma_{\underline{C}}$.

2.2.5 Proposition Chy is bireflective in Fil (and hence also in C). If (X,γ) is a filterspace then the Chy reflection is

$$1_X: (X,\gamma) \to (X,\underline{C}_\gamma)$$

where $\underline{C}_\gamma$ is the Cauchy structure defined by

> $\underline{F}$ is a Cauchy filter in $(X,\underline{C}_\gamma)$ if and only if there exists a finite sequence $\underline{F}_1,\ldots,\underline{F}_n$ of Cauchy filters in (X,γ) such that every member of $\underline{F}_i$ intersects every member of F_{i+1} for all $i < n$ and such that $\bigcap_{i=1}^{n} \underline{F}_i < \underline{F}$.

It follows that initial structures in Chy are constructed the same way as in Fil and that the final structures in Chy are constructed by applying the Chy bireflection to the final structure constructed in Fil.

2.2.6 Proposition

$Conv_R$ is a bicoreflective subcategory of Chy.
$Conv_S$ is a bicoreflective subcategory of C.
$Conv_R$ is a bireflective subcategory of $Conv_S$.

Proof: The embedding of $Conv_R$ as a coreflective subcategory of Chy was described in Proposition 1.3.5. Using the same procedure, $Conv_S$ is embedded bicoreflectively in C. With different terminology, details can be found in [104, p. 51]. The bireflector from $Conv_S$ to $Conv_R$ is the restriction of the bireflector from Fil to Chy described above.

2.2.7 Proposition $Pretop_R$ is bireflective in $Conv_R$.

Proof: Let ψ : Conv $\rightarrow$ Pretop be the reflector defined in Proposition 1.2.6. Then ψ preserves the symmetry axiom. So let us denote its restriction $Conv_S \rightarrow Pretop_S$ again by ψ.

Let σ be the reflector $Conv_S \rightarrow Conv_R$ described in Proposition 2.2.5. For X in $Conv_R$, let

$$X_1 = \sigma\psi X$$

$$X_{\alpha+1} = \sigma\psi X_\alpha \qquad \alpha \text{ an ordinal}$$

$$X_\alpha = \inf_{\beta<\alpha} X_\beta \qquad \alpha \text{ a limit ordinal}$$

Then there exists an ordinal γ such that 1_X: $X \rightarrow X_\gamma$ is the $Pretop_R$ reflection of X.

2.2.8 Proposition Top_R is bireflective in $Pretop_R$.

Proof: Let N: Mer $\rightarrow$ Near be the Near reflector as described in [58] and take the restriction of N to $Conv_S$. The reflector N when applied to a symmetric convergence space gives a symmetric topological space. Again, let σ : $Conv_S \rightarrow Conv_R$ be the reflector as described in Proposition 2.2.5. For X in $Pretop_R$ let

$$X_1 = \sigma NX$$

$$X_{\alpha+1} = \sigma N X_\alpha \qquad \alpha \text{ an ordinal}$$

$$X_\alpha = \inf_{\beta<\alpha} X_\beta \qquad \alpha \text{ a limit ordinal}$$

Then there exists an ordinal γ such that $1_X\colon X \to X_\gamma$ is the Top_R reflection of X.

We remark that the bireflector λ: Conv $\to$ Top as described in Proposition 1.2.8 could not be used in the proof of the previous proposition. λ does not preserve symmetry, so it cannot be composed with the reflector σ. Other reflections or coreflections in the diagram above are well known and can be found, for instance, in [57] and [104].

Combining the results about Chy obtained in Section 1.3 and those about Near from [104] we can now conclude this section with the following relations between the subcategories in our diagram.

2.2.9 Proposition

1. $Conv_R$ = Chy $\cap$ $Conv_S$
2. $Pstop_R$ = Chy $\cap$ $Pstop_S$
3. $Pretop_R$ = Chy $\cap$ $Pretop_S$
4. Top_R = Chy $\cap$ Top_S = Near $\cap$ $Conv_R$ = Near $\cap$ $Pretop_R$
5. Top_S = Near $\cap$ $Conv_S$ = Near $\cap$ $Pretop_S$

2.3 SEPARATED MEROTOPIC SPACES

2.3.1 Definition If X is a merotopic space, then a collection $\underline{A}$ of subsets of X is said to be concentrated provided that $\underline{A}$ is near and $\underline{A}$ is micromeric. X is said to be separated if for every concentrated collection $\underline{A}$, $\{B \subset X \mid \{B\} \cup \underline{A} \text{ is near}\}$, is near.

Sep denotes the full subcategory of Mer whose objects are the separated spaces.

Every regular nearness space [57] and a fortiori every uniform space is separated, and a T_1 topological space is separated if and only if it is Hausdorff [57]. Many of the proofs of results in [11] about Sep ∩ Near actually work as well for Sep. For example, Sep is productive and hereditary in Mer; therefore, since every indiscrete merotopic space belongs to Sep, Sep is bireflective in Mer.

In this section we will show that in the separated case our diagram of Section 2.2 reduces to the following one.

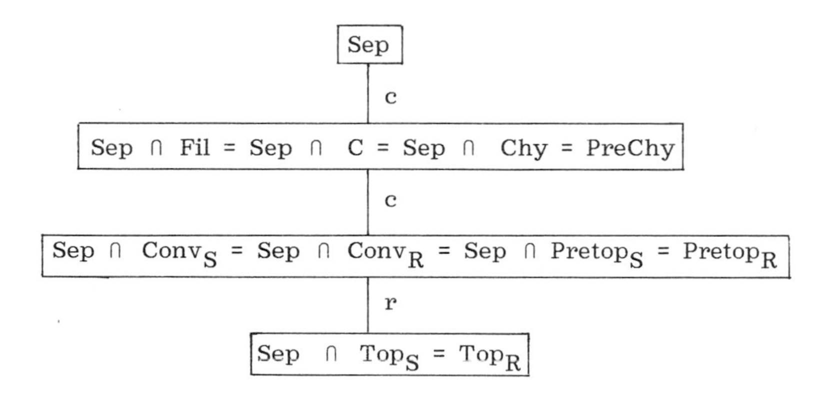

The condition that X be separated has an equivalent formulation in terms of Cauchy filters [12]. When we apply this formulation to a filterspace then we obtain the following characterization.

2.3.2 Proposition For a filter space X the following are equivalent:

1. X is separated.
2. Every Cauchy filter contains a unique minimal Cauchy filter.
3. Every Cauchy filter contains a smallest Cauchy filter.

Since in a Cauchy space X every Cauchy filter contains at most one minimal Cauchy filter we obtain the following characterizations.

2.3.3 Proposition For a Cauchy space X the following are equivalent:

1. X is separated.
2. Every Cauchy filter contains a minimal Cauchy filter.
3. Every Cauchy filter contains a smallest Cauchy filter.
4. Every equivalence class contains a minimal Cauchy filter.
5. X is a pretopological Cauchy space.

2.3.4 Proposition Every separated filter space is a Cauchy space.

Proof: Let $\underline{F}$ and $\underline{G}$ be Cauchy filters such that sup $(\underline{F},\underline{G})$ exists. Let $\underline{U}$ be an ultrafilter with sup $(\underline{F},\underline{G}) < \underline{U}$. Then $\underline{U}$ is a Cauchy filter and so by Proposition 2.3.2, $\underline{F}$ and $\underline{G}$ and $\underline{U}$ each contain a smallest Cauchy filter. It follows that there is a minimal Cauchy filter $\underline{H} < \underline{F} \cap \underline{G}$. Therefore $\underline{F} \cap \underline{G}$ is Cauchy.

2.3.5 Corollary

$$\text{Sep} \cap \text{Fil} = \text{Sep} \cap \text{C} = \text{Sep} \cap \text{Chy} = \text{PreChy}$$

The coreflector Mer $\rightarrow$ Fil preserves the property of being separated. Hence Sep $\cap$ Fil is bicoreflective in Sep. The coreflector C $\rightarrow$ Conv_S also preserves the property of being separated. Hence Sep $\cap$ Conv_S is bicoreflective in Sep $\cap$ C. Together with the results in Section 1.3 we now also have the following.

2.3.6 Corollary

1. Sep $\cap$ $Conv_S$ = Sep $\cap$ $Conv_R$ = Sep $\cap$ $Pretop_S$ = $Pretop_R$
2. Sep $\cap$ Top_S = Top_R

To describe the relation between topological Cauchy spaces introduced in Section 1.3 and the spaces in the categories in our diagram, we need some further properties of the underlying closure of a merotopic space. Let (X,γ) be a space in C. Then the underlying closure cl_γ is defined as in Definition 2.1.6. Let (X,q_γ) be the $Conv_S$ coreflection of (X,γ). Then closure in this convergence structure in the sense of Definition 1.2.5 is denoted by $cl_{q\gamma}$. The notations int_γ and $int_{q\gamma}$ have an analogous meaning.

2.3.7 Proposition For a space (X,γ) in C and for any $A \subset X$ we have

$$cl_{q_\gamma} A = cl_\gamma A$$

Proof: Suppose $x \in cl_{q\gamma} A$. Let $\underline{F}$ be a filter on X converging to x in q_γ and containing A. Then clearly

$$\underline{F} \cap \dot{x} < \sec \{A,\{x\}\}$$

It follows that $\sec \{A,\{x\}\} \in \gamma$.

Conversely, suppose $x \in cl_\gamma A$. Since γ is a filter structure we can find a Cauchy filter $\underline{F}$ in γ such that

$$\underline{F} < \sec \{A, \{x\}\}$$

It follows that sup $(\underline{F}$,stack A) exists and that moreover, sup $(\underline{F}$,stack A) $\cap \dot{x} > \underline{F} \cap \dot{x} = \underline{F}$. It then follows that sup $(\underline{F}$,stack A) converges to x in the convergence space (X,q_γ). Hence $x \in cl_{q\gamma}A$.

2.3.8 Corollary For a space (X, γ) in C and for any $A \subset X$ we have

$$\text{int}_\gamma A = \text{int}_{q_\gamma} A$$

2.3.9 Proposition

$$\text{Sep} \cap \text{Fil} \cap \text{Near} = \text{Sep} \cap \text{Chy} \cap \text{Near} = \text{TopChy}$$

Proof: Suppose (X, γ) is a topological Cauchy space. In view of Corollary 2.3.5 we already have that (X, γ) is a separated filter space. Now we show that (X, γ) satisfies the nearness axiom. We use the characterization by means of uniform covers. Let $\underline{A}$ be a uniform cover. Let $\underline{B} \in \gamma$ be arbitrary. Then $\underline{B}$ can be corefined by a Cauchy filter $\underline{F}$. Let $\underline{M}$ be the "open" base of the minimum Cauchy filter of the equivalence class of $\underline{F}$. Here the term "open" is used in the sense that for $M \in \underline{M}$ we have:

$$M = \text{int}_{q_\gamma} M$$

Since $\underline{A}$ is a uniform cover we can find an $A \in \underline{A}$ and an $M \in \underline{M}$ with $M \subset A$. In view of the preceding proposition we now have

$$\text{int}_\gamma A = \text{int}_{q_\gamma} A \supset \text{int}_{q_\gamma} M = M$$

It follows that

$$\{\text{int}_\gamma A \mid A \in \underline{A}\} \cap \text{stack } \underline{B} \neq \emptyset$$

In view of the arbitrariness of $\underline{B}$ we can conclude that,

$$\{\text{int}_\gamma A \mid A \in A\} \text{ is a uniform cover}$$

Conversely, suppose (X, γ) is a separated Cauchy space satisfying the nearness axiom. By Corollary 2.3.5, we already have that (X, γ) is a pretopological Cauchy space. So it suffices to show that for every Cauchy equivalence class the minimum Cauchy filter has an "open" base where for $M \subset X$, "M open" means $\mathrm{int}_{q_\gamma} M = M$. Suppose $\underline{M}$ is such a minimum Cauchy filter. Let $\underline{A}$ be a uniform cover of (X,γ). Since (X,γ) satisfies the nearness axiom, there exists $A \in \underline{A}$ such that $\mathrm{int}_\gamma A \in \underline{M}$. Since int_γ is idempotent and $\mathrm{int}_\gamma = \mathrm{int}_{q_\gamma}$ it follows that $\mathrm{int}_\gamma A$ is q_γ-open. Since $A \supset \mathrm{int}_\gamma A$ we now can conclude that

$$A \in \underline{A} \cap \text{stack } \{M \in \underline{M} \mid M \text{ is } q_\gamma\text{-open}\}$$

So $\{M \in \underline{M} \mid M \text{ is } q_\gamma\text{-open}\}$ is micromeric. It follows that it is an open base of $\underline{M}$.

2.4 SUBSPACES IN MER

From the results and the summarizing diagrams in Sections 2.2 and 2.3 we can conclude that the following subcategories of Mer,

(Fil, C, Chy, Near, Sep)

all have the same subspace structures as in Mer. Moreover we know that any Hausdorff topological space Y is an object of each of these five categories. So if X is a subset of Y, then the subspace structure on X, taken in any of these categories, results in the same merotopic space, which then again belongs to each of these five categories.

In this section we shall give a precise characterization of the merotopic subspaces of Hausdorff topological spaces and we shall compare them to the merotopic subspaces of more restricted spaces, such as completely uniformizable Hausdorff topological spaces on one hand, and to merotopic subspaces of more general spaces, such as Hausdorff convergence spaces, on the other. Subspaces of reciprocal or symmetric convergence spaces are studied in [14]. To deal with the Hausdorff case we first have to recall the following separation condition for merotopic spaces.

2.4.1 Definition A merotopic space is said to be T_1 if it satisfies the condition

$$\dot{x} \cap \dot{y} \text{ is micromeric implies } x = y$$

We remark that for Cauchy spaces and for symmetric convergence spaces this definition coincides with the T_1 properties introduced in Definition 1.3.10 and Definition 1.2.9 respectively.

Let T_1 denote the full subcategory of Mer, whose objects are the T_1 merotopic spaces. Further, let

$$T_1 \text{ Fil} = T_1 \cap \text{Fil}$$

The categories T_1C, T_1Chy, T_1Conv, T_1Near, $T_1\text{Chy}_U$, T_1PsChy, T_1PreChy, and T_1TopChy are all defined analogously.

2.4.2 Proposition Subspaces of T_1 merotopic spaces are T_1.

The following notation will be useful in the formulation of the results.

2.4.3 Definition If $\underline{A}$ is a full subcategory of Mer, then sub $\underline{A}$ will denote that full subcategory of Mer whose objects are all merotopic subspaces of spaces in $\underline{A}$.

2.4.4 Theorem

1. sub $\text{Conv}_H = T_1\text{Chy}$
2. sub $\text{Pstop}_h = T_1\text{PsChy}$

3. sub $Pretop_H$ = $T_1Chy \cap Sep$ = $T_1PreChy$
4. sub Haus = $T_1Chy \cap Near \cap Sep$ = $T_1TopChy$
5. sub $Haus_{CU}$ = T_1Chy_U
6. sub $Conv_{H.Comp}$ = category of T_1 totally bounded Cauchy spaces
7. sub $Conv_{c\text{-}emb}$ = category of c^-embedded Cauchy spaces

Proof: 1. $Conv_H \subset T_1Chy$ and T_1Chy is hereditary in Mer. Hence sub $Conv_H \subset T_1Chy$.

To show the reverse inclusion, let X be a T_1Cauchy space. Then let Y be any Hausdorff completion of X. As was done before, Y then can be identified with a Hausdorff convergence space in which X is a merotopic subspace.

2. $Pstop_H \subset T_1PsChy$ and T_1PsChy is hereditary in Mer. Hence sub $Pstop_H \subset T_1PsChy$.

To show the other inclusion, let X be a T_1 pseudotopological Cauchy space, and let Y be any completion in the class $\underline{R}$. Then Y is a Hausdorff pseudotopological convergence space in which X is a merotopic subspace.

3. $Pretop_H \subset T_1Chy \cap Sep$ and T_1Chy and Sep are hereditary in Mer, so sub $Pretop_H \subset T_1Chy \cap Sep$.

To show the reverse inclusion, let X be a space in $T_1Chy \cap Sep$. Then X is a pretopological T_1 Cauchy space. Take for Y the Wyler completion of X, or the Kowalsky completion, or any completion of the type $K_\omega(X)$ as described in Proposition 1.4.11. Then Y is a pretopological Hausdorff convergence space containing X as a merotopic subspace.

4. Haus $\subset Pretop_H \subset T_1Chy \cap Sep$ and Haus $\subset$ Near. Moreover T_1Chy, Sep, and Near are all hereditary in Mer. This gives one inclusion.

For the other inclusion, observe that a space X in $T_1Chy \cap Sep \cap Near$ is a topological Cauchy space and that the Wyler completion (or the Kowalsky completion) then is a Hausdorff topological space containing X as a merotopic subspace.

5. Haus $\subset$ T_1Chy_U. Moreover T_1Chy_U is hereditary in Mer, from the following observations: If $(Y,\underline{U})$ is a uniform space and $X \subset Y$, then we can either first take the merotopic subspace on X and then take the filter coreflection, or we can first take the filter coreflection $(Y,\underline{C}_{\underline{U}})$ and then take the merotopic subspace on X. Since Fil is hereditary in Mer, it follows that in both cases we end up with the same merotopic structure. It follows that the subspace of $(Y,\underline{C}_{\underline{U}})$ is uniformizable. So we can conclude that sub $\text{Haus}_{CU} \subset \overline{T}_1Chy_U$.

To show the reverse inclusion let X be a T_1 uniformizable Cauchy space. The Kowalsky completion then is a completely uniformizable topological space containing X as a merotopic subspace.

6. Every compact Hausdorff convergence space is a T_1 totally bounded Cauchy space. Moreover T_1 totally bounded Cauchy spaces are hereditary in Mer. This shows one inclusion.

For the other inclusion observe that when X is a T_1 totally bounded Cauchy space, any completion of the type $K_\omega(X)$ is a compact Hausdorff convergence space containing X as a merotopic subspace.

7. Every c-embedded convergence space is a $c\hat{}$-embedded Cauchy space, and $c\hat{}$-embedded Cauchy spaces are hereditary in Mer. This gives one inclusion.

To show the other inclusion let X be any $c\hat{}$-embedded Cauchy space. Then the natural completion $\underline{N}(X)$ Proposition 1.4.15 is a c-embedded convergence space containing X as a merotopic subspace.

3

Classes of Real-Valued Continuously Extendible Functions

In this chapter we come to the main subject of our work, namely, the study of the function classes of Cauchy continuous maps. We start with an arbitrary T_1 Cauchy space X and consider the class $\Gamma(X)$ of all Cauchy continuous maps from X to the real line with the usual topology. First we take a strict completion Y of X and the bijection κ mapping a function f to its extension $\hat{f}$:

$$\kappa : \Gamma(X) \to C(Y) : f \to \kappa(f) = \hat{f}$$

C(Y) is an algebra and a lattice containing the constants, C(Y) is strongly composition closed, uniformly complete, and complete for continuous convergence. In Section 3.2 we investigate whether these properties of C(Y) can be carried over to $\Gamma(X)$ via κ. We prove that $\Gamma(X)$ too is an algebra and a lattice containing the constants, and that $\Gamma(X)$ is composition closed and uniformly complete. We give examples showing that strongly composition closedness and completeness for continuous convergence, however, are not carried over to $\Gamma(X)$. Now $\Gamma^*(X)$, the function class of all bounded functions in $\Gamma(X)$, is equal to the class of all Cauchy continuous maps on the totally bounded reflection of X. This result is established in Section 3.3 and implies that $\Gamma^*(X)$ shares all the nice properties of $\Gamma(X)$.

In the last two sections of this chapter we investigate "domains." These are Cauchy spaces that are equal to the initial Cauchy structure determined by their class of Cauchy continuous maps. We characterize realcompact and compact topological spaces as complete domains and complete totally bounded domains, respectively.

3.1 REAL-VALUED MORPHISMS ON A FILTER MEROTOPIC SPACE

Properties of the function class of all morphisms on a merotopic space X to the real line will of course depend on

1. The type of merotopic structure on the domain X.
2. The choice of a natural merotopic structure on the real line.

For example, the class of all functions from $\mathbf{R}$ to $\mathbf{R}$, uniformly continuous in the usual sense, does not have nice properties, and is not even a function algebra, whereas the class of all functions from $\mathbf{R}$ to $\mathbf{R}$, continuous in the usual sense, is an inversion closed function algebra with many other nice features.

H. Bentley, M. Hastings, and R. Ori [8] restricted themselves to locally fine nearness spaces as domain spaces and showed that the function class of all real-valued morphisms on a locally fine nearness space is a Φ-algebra in the sense of [54], and has other nice properties. In this section we start by restricting ourselves to filter merotopic spaces as domain spaces. So our domain spaces are locally fine spaces but not necessarily nearness spaces.

On the real line we will consider two different natural merotopies. As in [10], R with the merotopy of the usual topology is denoted by R_t. The merotopic structure of R_t coincides with the complete Cauchy structure of the usual uniformity. On the other hand the usual uniformity itself also makes the real line into a merotopic space. This space will be denoted by R_u. Both structures R_t and R_u are different, and moreover R_t is the filter coreflection of R_u.

For a merotopic space (X,γ) we denote by $\Gamma((X,\gamma),R_t)$ and $\Gamma((X,\gamma),R_u)$ the collections of all morphisms in Mer (i.e., uniformly continuous functions) from (X,γ) to the real line, endowed with R_t and R_u, respectively.

3.1.1 Proposition If (X,γ) is a filter merotopic space and $(X,\underline{C}_\gamma)$ is its Cauchy reflection, then we have

$$\Gamma((X,\gamma)\ R_t) = \Gamma((X,\underline{C}_\gamma),\ R_t) = \Gamma((X,\underline{C}_\gamma),\ R_u)$$
$$= \Gamma((X,\gamma),\ R_u)$$

Proof: Since R_t is the filter coreflection of R_u and since (X,γ) is a filter space clearly we have the equality

$$\Gamma((X,\gamma),\ R_t) = \Gamma((X,\gamma),\ R_u)$$

For the same reason we also have the equality

$$\Gamma((X,\underline{C}_\gamma),\ R_t) = \Gamma((X,\underline{C}_\gamma),\ R_u)$$

So it suffices to show that $\Gamma((X,\gamma),\ R_t)$ and $\Gamma((X,\underline{C}_\gamma),\ R_t)$ are equal too. This follows at once, since $(X,\underline{C}_\gamma)$ is the Cauchy reflection of (X,γ) and since R_t is a Cauchy space.

From the preceding proposition it becomes clear that for the study of algebraic properties of function classes of real-valued morphisms on a filter merotopic space X, and without any loss of generality, we can make the additional assumptions that the filter space X is a Cauchy space, and that the real line carries the structure R_t. The structure R_u will be used only in Section 3.4, where the function class is endowed with uniform convergence.

For the rest of this chapter, $(X,\underline{C})$ will be a T_1 Cauchy space and when no confusion can arise, the class $\Gamma((X,\underline{C}), \mathsf{R}_t)$ will be denoted by $\Gamma(X,\underline{C})$ or simply by $\Gamma(X)$. This function class will be called the function class of real-valued Cauchy continuous maps on X.

The results of the first chapter moreover imply that in studying these function classes $\Gamma(X)$ of real-valued Cauchy continuous maps, we capture all the examples of function classes arising as restriction classes $C(Y)/_X$ of classes of continuous maps on strict Hausdorff extensions of X. In particular, we capture all the function classes C(X).

The completion theory developed in the first chapter further implies that, for every class $\Gamma(X)$ there exists an extension Y of X such that $C(Y)/_X = \Gamma(X)$. In this context, the extension Y is not uniquely determined. Uniqueness of Y can be obtained only in a more restrictive situation. An example of such a situation, where uniqueness of Y is indeed obtained, is developed in Chapter 4.

3.2 COMPOSITION PROPERTIES OF $\Gamma(X)$

In this section X is a T_1 Cauchy space and $\Gamma(X)$ is its collection of real-valued Cauchy continuous maps. All extensions are assumed to be Hausdorff extensions.

3.2.1 Proposition $\Gamma(X)$ is a commutative function algebra over R and a lattice for the pointwise operations. Moreover $\Gamma(X)$ contains the constants.

Proof: Let Y be any strict completion of X. For f,g in $\Gamma(X)$ let $\hat{f},\hat{g}$ be the extensions belonging to C(Y) as constructed in Theorem 1.4.12. Since all operations on $\Gamma(X)$ are defined pointwise and since C(Y) is an algebra over **R** and a lattice we clearly have:

$$\hat{f} + \hat{g}\,|_X = f + g \in \Gamma(X); \; \hat{f} \cdot \hat{g}\,|_X = f \cdot g \in \Gamma(X)$$

$$\sup(\hat{f},\hat{g})\,|_X = \sup(f,g) \in \Gamma(X); \; \inf(\hat{f},\hat{g})\,|_X = \inf(f,g) \in \Gamma(X)$$

$$-\hat{f}\,|_X = -f \in \Gamma(X); \; \hat{1}\,|_X = 1 \in \Gamma(X)$$

In what follows we denote by κ the map

$$\kappa: \; \Gamma(X) \to C(Y): \; f \to \kappa(f) = \hat{f}$$

3.2.2 Proposition κ is an algebra and a lattice isomorphism.

Proof: This follows immediately from the uniqueness of the extensions.

We shall now show that $\Gamma(X)$ also satisfies some stronger composition properties than those from Proposition 3.2.1. The following notions were introduced by Császár in [31].

3.2.3 Definition A function class $\Phi \subset \mathbf{R}^X$ is <u>composition closed</u> <u>(strongly composition closed)</u> if the following condition is satisfied: Let $\{ f_i \mid i \in I \} \subset \Phi$ and let h be the map defined by h: $X \to \mathbf{R}^I$: $x \to h(x) = (f_i(x))_{i \in I}$.

For every real-valued continuous map k defined on the closure of h(X) in $\mathbf{R}^I$ (defined on h(X)), the map $k \circ h$ belongs to Φ.

3.2.4 Proposition $\Gamma(X)$ is composition closed.

Proof: Let $\{f_i \mid i \in I\} \subset \Gamma(X)$, and let h and k be defined as in Definition 3.2.3. We show that $k \circ h$ belongs to $\Gamma(X)$. If $\underline{F}$ is a Cauchy filter then stack $h(\underline{F})$ converges in $\mathbf{R}^I$ and furthermore, since $h(X) \in h(\underline{F})$, the filter stack $h(\underline{F})$ also converges on cl $h(X)$. Consequently stack $k(h(\underline{F}))$ converges in $\mathbf{R}$, which by the arbitrariness of $\underline{F}$ implies that $k \circ h \in \Gamma(X)$.

$\Gamma(X)$ is in general not strongly composition closed. When $f \in \Gamma(X)$ is never zero on X and $k = 1/x$ is defined on $f(X)$ then the composition $k \circ h = 1/f$ is not always Cauchy continuous. Take, for instance, $X = \mathbf{R}\setminus\{0\}$ endowed with the Cauchy structure induced by the extension $\mathbf{R}$ and take $f: \mathbf{R}\setminus\{0\} \to \mathbf{R}: f(x) = x$. Then the inverse $1/f$ is clearly not Cauchy continuous. We shall establish necessary and sufficient conditions for a Cauchy continuous function to have a Cauchy continuous inverse.

3.2.5 Lemma If Y is any strict completion of X then a function in $\Gamma(X)$ is invertible in $\Gamma(X)$ if and only if the continuous extension $\hat{f}$ is never zero on Y.

Proof: The "if" part follows from $(1/\hat{f})\mid_X = 1/f$. To prove the "only if " part if suffices to remark that, with the notation of Proposition 3.2.1

$$1 = \kappa(1) = \kappa(f \cdot \frac{1}{f}) = \kappa(f) \cdot \kappa(\frac{1}{f})$$

It follows that $\kappa(f)$ is invertible in $C(Y)$.

The result in Lemma 3.2.5 implies that f is invertible in $\Gamma(X)$ if and only if f^2 is invertible in $\Gamma(X)$. So it suffices to formulate conditions on the invertibility of positive functions.

3.2.6 Definition Let A be a subset of X. A function $f \in \mathbf{R}^X$ is <u>bounded away from zero on A</u> if

$$\inf_{x \in A} |f(x)| > 0$$

3.2.7 Definition A cover $\underline{A}$ of a Cauchy space X is a <u>Cauchy cover</u> if the following condition holds:

for every Cauchy filter $\underline{F}$: $\underline{F} \cap \underline{A} \neq \emptyset$

It is clear that a cover $\underline{A}$ of a Cauchy space is a Cauchy cover if and only if $\underline{A}$ is a uniform cover in the sense of Definition 2.1.8.

3.2.8 Proposition A strictly positive function f in $\Gamma(X)$ is invertible in $\Gamma(X)$ if and only if X has a Cauchy cover $\underline{A}$ such that on every $A \in \underline{A}$ the function f is bounded away from zero.

Proof: To show the "only if" part, suppose f is a strictly positive function in $\Gamma(X)$ and f is invertible in $\Gamma(X)$. Let Y be any strict completion of X. Applying Lemma 3.2.5 we have that $\hat{f}$ is never zero on Y. If $\underline{F}$ is a Cauchy filter we have lim stack $f(\underline{F}) = a > 0$. Take $0 < \varepsilon < a$ and choose $F \in \underline{F}$ such that

$$f(F) \subset (a - \varepsilon, a + \varepsilon)$$

Clearly, then f is bounded away from zero on F.

Conversely, to show the "if" part, suppose $\underline{A}$ is a Cauchy cover and f is bounded away from zero on every $A \in \underline{A}$. For a Cauchy filter $\underline{F}$ we choose $A \in \underline{A} \cap \underline{F}$. For this set we have

$$0 \notin \mathrm{cl}_{\mathbf{R}} f(A)$$

and for the Cauchy filter $\underline{F}$ this implies

$\lim \text{stack } f(\underline{F}) \neq 0$

Now if y is any strict completion of X and $\hat{f}$ is the extension of f then $\hat{f}$ is never zero on Y and f is invertible in $\Gamma(X)$ in view of Lemma 3.2.5.

The following notions are well known from the literature on function classes. However, the terminology is not uniform. The terminology we use for conditions (1) and (3) conforms to the papers of Isbell [61] or Hendriksen and Johnson [54]. Condition (2) was introduced by Mrówka [94]. However Mrówka uses the term "bounded inversion closed" for (2).

3.2.9 Definition Let $\Phi \subset \mathbf{R}^X$ be a function class.

1. Φ is <u>bounded inversion closed</u> if $f \in \Phi$, f bounded away from 0, implies $1/f \in \Phi$.
2. Φ is <u>strongly bounded inversion closed</u> if $f, g \in \Phi$, f never zero on X, g/f bounded, implies $g/f \in \Phi$.
3. Φ is <u>inversion closed</u> if $f \in \Phi$, f never zero on X, implies $1/f \in \Phi$.

From Proposition 3.2.8 we immediately have the next result.

3.2.10 Proposition $\Gamma(X)$ is bounded inversion closed.

3.2.11 Remark Many examples occur where $\Gamma(X)$ is not inversion closed. We already mentioned the example of $X = \mathbf{R}\setminus\{0\}$ endowed with the Cauchy structure induced by its extension $\mathbf{R}$ and $f: X \to \mathbf{R}$, $f(x) = x$, which has no inverse in $\Gamma(X)$.

Even the weaker condition (2) of strongly bounded inversion closedness does not hold in this example. Take $f(x) = x$ and $g(x) = |x|$ on $\mathbf{R}\setminus\{0\}$. Then g/f is bounded but not Cauchy continuous.

Inversion closedness and strongly bounded inversion closedness illustrate the difference between a general collection $\Gamma(X)$ on one hand, and a particular collection $C(X)$ of continuous functions on the other. In Chapter 4 the effect of these two properties on the structure space theory of $\Gamma(X)$ is investigated. Also in Chapter 4, other differences between $\Gamma(X)$ and $C(X)$ will be illustrated. It will be shown that even inversion closed collections $\Gamma(X)$ need not be equal to some collection $C(X)$ of continuous maps.

3.3 BOUNDED FUNCTIONS

In this section X is again a T_1 Cauchy space and $\Gamma(X)$ is its collection of real-valued Cauchy continuous maps. Let $\Gamma^*(X)$ be the collection of bounded functions in $\Gamma(X)$. We shall now investigate composition properties of this smaller function class.

3.3.1 Lemma If X is totally bounded, then all functions in $\Gamma(X)$ are bounded.

Proof: Suppose X is totally bounded and let $f \in \Gamma(X)$. For every ultrafilter $\underline{W}$ on X we can find a compact set $K \subset \mathbf{R}$ and $W \in \underline{W}$ such that $f(W) \subset K$. Then a finite collection of ultrafilters $\underline{W}_1, \ldots, \underline{W}_n$, corresponding compact sets $K_1, \ldots, K_n$, and sets $W_1, \ldots, W_n$ with $W_i \in \underline{W}_i$ and $f(W_i) \subset K_i$ can be selected such that $W_1 \cup \cdots \cup W_n = X$. Finally we have

$$f(X) = f\left(\bigcup_{i=1}^{n} W_i\right) = \bigcup_{i=1}^{n} f(W_i) \subset \bigcup_{i=1}^{n} K_i$$

and hence f is bounded.

3.3.2 Remark If $\Gamma(X)$ consists of bounded functions, then X need not be totally bounded. For example, let $X = [0,1]$ and let $\underline{V}(x)$ be the usual neighborhoodfilter of $x \in X$. Fix a nonprincipal ultrafilter $\underline{U}_0 > \underline{V}(0)$. We make X a Cauchy space by defining a filter $\underline{F} \in F(X)$ to be a Cauchy filter if

$$\text{either } \underline{F} > \underline{V}(x) \text{ for some } x \neq 0$$
$$\text{or } \underline{F} > \underline{V}(0) \text{ and } \underline{U}_0 \not> \underline{F}$$

Clearly X is not totally bounded. On the other hand, however, for $f \in \Gamma(X)$, in the first place, if $x \neq 0$ we clearly have that stack $f(\underline{V}(x))$ converges to $f(x)$.

In the second place, if $x = 0$, we note that for two ultrafilters $\underline{W}_1$ and $\underline{W}_2$, both finer than $\underline{V}(0)$ and different from $\underline{U}_0$, we have

$$\lim \text{ stack } f(\underline{W}_1) = \lim \text{ stack } f(\underline{W}_2) = f(0)$$

Moreover

$$\begin{aligned} \text{stack } f(\underline{V}(0)) &= \text{stack } f(\cap\{\underline{W} \mid \underline{W} \in U(X),\ \underline{W} > \underline{V}(0),\\ &\qquad \underline{W} \neq \underline{U}_0\}) \\ &= \cap\{\text{stack } f(\underline{W}) \mid \underline{W} \in U(X),\ \underline{W} > \underline{V}(0),\\ &\qquad \underline{W} \neq \underline{U}_0\} \end{aligned}$$

So stack $f(\underline{V}(0))$ converges to $f(0)$. This shows that f is continuous for the usual topology on $[0,1]$ and hence f is bounded.

As a consequence of Lemma 3.3.1 we obtain the following permanence property for Cauchy continuous maps.

3.3.3 Corollary Every Cauchy continuous function on X maps bounded subsets of X to (totally) bounded subsets of **R**.

Proof: Let $f \in \Gamma(X)$ and let $A \subset X$ be a totally bounded subset. The restriction $f|_A$ of f to A is Cauchy continuous on A, endowed with the induced Cauchy structure. Since A is totally bounded, $f|_A$ is bounded. So $f(A)$ is a bounded subset of **R**.

The proof of the following lemma is straightforward.

3.3.4 Lemma For any Cauchy space $(X,\underline{C})$ the collection

$$\underline{C}_t = \underline{C} \cup U(X)$$

is a totally bounded Cauchy structure on X. Moreover the identity mapping $1_X:(X,\underline{C}) \to (X,\underline{C}_t)$ is the reflection of $(X,\underline{C})$ in the full subcategory of totally bounded Cauchy spaces.

3.3.5 Proposition For any Cauchy space X we have

$$\Gamma^*(X,\underline{C}) = \Gamma(X,\underline{C}_t)$$

Proof: Since $\underline{C}_t \leq \underline{C}$ we have $\Gamma(X,\underline{C}_t) \subset \Gamma(X,\underline{C})$ and since $\underline{C}_t$ is totally bounded we can apply Lemma 3.3.1. So we have $\Gamma(X,\underline{C}_t) \subset \Gamma^*(X,\underline{C})$.

Conversely, if $f \in \Gamma^*(X,\underline{C})$ and $\underline{F}$ is a Cauchy filter for $\underline{C}_t$, then either $\underline{F}$ is also a Cauchy filter for $\underline{C}$ and then stack $f(\underline{F})$ converges, or $\underline{F} \in U(X)$ and then stack $f(\underline{F})$ is an ultrafilter on the bounded set $f(X)$ and again stack $f(\underline{F})$ converges.

3.3.6 Corollary $\Gamma^*(X)$ is a commutative function algebra over R and a lattice for the pointwise operations, containing the constants. Moreover $\Gamma^*(X)$ is composition closed.

3.3.7 Proposition A strictly positive function f in $\Gamma^*(X)$ is invertible in $\Gamma^*(X)$ if and only if f is bounded away from zero.

Proof: If f is bounded away from zero then it suffices to apply Proposition 3.2.8 to $\Gamma(X,\underline{C}_t)$ and we are done.

For the other implication, suppose $f > 0$ is invertible in $\Gamma^*(X)$. Again, by applying Proposition 3.2.8 to $\Gamma(X,\underline{C}_t)$ we have a Cauchy cover $\underline{A}$ for $\underline{C}_t$ such that f is bounded away from zero on every $A \in \underline{A}$. For every $\underline{U} \in U(X)$, choose $A \in \underline{A} \cap \underline{U}$ with $\inf f(A) > 0$. We can select a finite number of ultrafilters $\underline{U}_1,\ldots,\underline{U}_n$ and corresponding sets $A_1,\ldots,A_n$ with $A_i \in \underline{A} \cap \underline{U}_i$ such that $X = A_1 \cup \cdots \cup A_n$. It follows that $\inf f(X) > 0$.

3.3.8 Remark Results comparable to (3.2.1), (3.2.8), (3.2.10), (3.3.5) and (3.3.7) were obtained earlier by Bentley, Hastings, and Ori in [8] for the collections of morphisms on locally fine nearness spaces. These merotopic spaces are not necessarily filter spaces and the function class of all real-valued morphisms on a locally fine nearness space does not share all nice properties with the classes $\Gamma(X)$; for instance, it is in general not composition closed.

3.4 $\Gamma(X)$ AS A FUNCTION SPACE

Let $(X,\underline{C})$ be a T_1 Cauchy space, and as in Proposition 1.3.5 let $(X,q_{\underline{C}})$ be the convergence coreflection. We put $\Gamma(X) = \Gamma(X,\underline{C})$ and $C(X) = C(X,q_{\underline{C}})$. Then clearly $\Gamma(X)$ is a subset of $C(X)$ and we let $i: \Gamma(X) \to C(X)$ be the canonical injection. Then $\Gamma(X)$ inherits uniform convergence, pointwise convergence, and continuous convergence from $C(X)$. $\Gamma(X)$ will be endowed with these induced structures.

Let Y be any strict Hausdorff completion of $(X,\underline{C})$, let $C(Y)$ be the class of real-valued continuous function on Y, and let $\kappa: \Gamma(X) \to C(Y)$ be defined as in Proposition 3.2.1. As we have seen, κ is an algebraic isomorphism. We shall now investigate whether the uniform convergence, pointwise convergence, or continuous convergence on $\Gamma(X)$ and $C(Y)$ makes κ also a "topological" isomorphism.

3.4.1 Proposition When $\Gamma(X)$ and $C(Y)$ are endowed with the uniformities of uniform convergence then $\kappa : \Gamma(X) \to C(Y)$ is a uniform isomorphism.

Proof: If V is any entourage for the usual uniformity on $\mathbf{R}$, choose V' an entourage such that $V'^3 \subset V$. Suppose $f \in \Gamma(X)$, $g \in \Gamma(X)$ are such that $(f(x), g(x)) \in V'$ for every $x \in X$. For $y \in Y$ choose a Cauchy filter $\underline{F}$ such that stack $\underline{F}$ converges to y in Y. Then $cl_{\mathbf{R}} f(\underline{F})$ and $cl_{\mathbf{R}} g(\underline{F})$ are Cauchy filters on $\mathbf{R}$. Choose $F \in \underline{F}$ such that:

$$cl_{\mathbf{R}} f(F) \times cl_{\mathbf{R}} f(F) \subset V'$$

and

$$cl_{\mathbf{R}} g(F) \times cl_{\mathbf{R}} g(F) \subset V'$$

Further, choose $x \in F$. Then we have

$$(\hat{f}(y), f(x)) \in V', \ (f(x), g(x)) \in V' \quad \text{and}$$

$$(g(x), \hat{g}(y)) \in V'$$

so that finally $(\hat{f}(y), \hat{g}(y)) \in V$. This proves that κ is uniformly continuous. That κ^{-1} also is uniformly continuous, is an easy verification.

The fact that $C(Y)$ is complete for uniform convergence and that κ is a uniform isomorphism implies the following result.

3.4.2 Corollary $\Gamma(X)$ is uniformly closed in $C(X)$.

Denseness of $\Gamma(X)$ in $C(X)$, which in view of the last result comes down to the equality of $\Gamma(X)$ and $C(X)$, is treated in Chapter 4.

For the structure of pointwise convergence the results are quite different. The proof of the first result is easy and we leave it to the reader.

3.4.3 Proposition If $\Gamma(X)$ and $C(Y)$ are endowed with the uniformities of pointwise convergence, then $\kappa^{-1}: C(Y) \to \Gamma(X)$ is uniformly continuous.

3.4.4 Remark When $\Gamma(X)$ and $C(Y)$ have the topology of pointwise convergence, κ need not be continuous. For example, let $X = (0,1]$ with the Cauchy structure induced by $[0,1]$. For $n > 0$ let $f_n: (0,1] \to \mathbf{R}$ be the function defined by

$$f_n(x) = \begin{cases} 0 & \text{for } x \geq \frac{1}{n} \\ n - n^2 x & \text{for } x < \frac{1}{n} \end{cases}$$

Then clearly the functions f_n are Cauchy continuous and converge pointwise to $f = 0$. However, when the functions are extended to $[0,1]$ the sequence $(\hat{f}_n)_{n>0}$ does not converge pointwise to $\hat{f}$.

3.4.5 Remark In general $\Gamma(X)$ is not pointwise closed in $C(X)$. On $(0,1]$ for example, every continuous function is the pointwise limit of a sequence of Cauchy continuous functions. For if h is continuous on $(0,1]$, then such a sequence can easily be defined as follows:

$$f_n(x) = \begin{cases} h(x) & \text{for } x \geq \frac{1}{n} \\ h(\frac{1}{n}) & \text{for } x < \frac{1}{n} \end{cases}$$

Then clearly $(f_n)_{n>0}$ converges pointwise to h. The same density results holds in other situations, for example, when the Cauchy space X is a subspace of a completely regular topological space. This will follow from the general situation treated in Proposition 3.4.10.

Next we look at the case of continuous convergence. The structure $\underline{G}_c$ of continuous convergence was introduced in [29]. It is a uniform convergence structure on C(X) defined as follows

$$\underline{G}_c = \{\Psi \in F(C(X) \times C(X)) \mid \Psi(\underline{F} \times \underline{F}) > \underline{V} \text{ for every convergent filter } \underline{F} \text{ on } X\}$$

where $\underline{V}$ is the usual uniformity on R.

Let $\underline{C}_{\underline{G}_c}$ be the Chy coreflection as described in Proposition 1.3.5

$$\underline{C}_{\underline{G}_c} = \{\Theta \in F(C(X)) \mid \Theta \times \Theta \in \underline{G}_c\}$$

Then the convergence coreflection of $\underline{C}_{\underline{G}_c}$ coincides with the continuous convergence on C(X). Moreover, on C(X) the Cauchy structure $\underline{C}_{\underline{G}_c}$ is complete. So on C(X) we have

$$\underline{C}_{\underline{G}_c} = \{\Theta \in F(C(X)) \mid \Theta \text{ converges continuously}\}$$

The following result is shown by straightforward verification.

3.4.6 Proposition When Γ(X) and C(Y) are endowed with the uniform convergence structure of continuous convergence then the map

$$\kappa^{-1} : C(Y) \to \Gamma(X)$$

is uniformly continuous.

3.4.7 Remark When $\Gamma(X)$ and $C(Y)$ have the continuous convergence structure, the map $\kappa : \Gamma(X) \to C(Y)$ need not be continuous. Take again the Example in (3.4.4). The sequence $(f_n)_{n>0}$ converges continuously to $f = 0$ since for $x \in (0,1]$ and $\underline{V}(x)$ the neighborhood filter in x we have

$$< f_n > (\underline{V}(x)) = \dot{0}$$

However when the functions are extended to [0,1] the sequence $(\hat{f}_n)_{n>0}$ does not converge continuously to $\hat{f}$.

If however we endow $\Gamma(X)$ with the Cauchy continuous Cauchy structure instead of with $\underline{C}_{G_c}$ then κ is an isomorphism. This is formulated in the next proposition, the proof of which can be found in [47].

3.4.8 Proposition When $\Gamma(X)$ is endowed with $\underline{\hat{C}}$ and $C(Y)$ is endowed with $\underline{C}_{G_c}$ then $\kappa : \Gamma(X) \to C(Y)$ is a Cauchy isomorphism.

We use the following notation. If $\Phi \subset \mathbb{R}^X$ then the initial topology on X determined by Φ is denoted by $t\Phi$. We say that Φ generates the topology τ if $t\Phi = \tau$.

In order to prove a density theorem, we give a formulation of Theorem 59 in [21] for the point-separating case.

3.4.9 Lemma Let X be a convergence space and $\underline{A} \subset C(X)$ a subalgebra containing the constants. If $\underline{A}^*$, the algebra of bounded functions in $\underline{A}$, is point separating and generates the topology $tC(X)$, then $\underline{A}$ is dense in $C(X)$ for continuous convergence.

3.4.10 Proposition If C(X) is point separating then any of the following properties implies the next one.

1. Γ(X) has the same initial topology as C(X).
2. Γ(X) is dense in C(X) for continuous convergence.
3. Γ(X) is dense in C(X) for pointwise convergence.
4. Γ(X) separates the points of X.

Proof:(1)⇒(2) Suppose tΓ(X) = t(C(X)). From Theorem 1.4 in Isbell's paper [61] and from Propositions 3.2.1 and 3.2.10 we can conclude that the collection Γ*(X) determines the same initial topology as Γ(X). So Γ*(X) is point separating and generates tC(X). Applying the previous Lemma we can conclude that Γ(X) is dense in C(X) for continuous convergence.

(2)⇒(3) This follows at once from the fact that pointwise convergence is coarser than continuous convergence.

(3)⇒(4) Suppose x and y are different points in X. First we choose a continuous function f separating x and y and then a filter Θ on C(X) containing Γ(X) and converging pointwise to f. Then clearly the filters stack $\Theta(\dot{x})$ and stack $\Theta(\dot{y})$ contain disjoint sets. We take $\underline{A} \in \Theta$ and $\underline{B} \in \Theta$ such that the intersection of $\underline{A}(x)$ and $\underline{B}(y)$ is empty and then choose $g \in \underline{A} \cap \underline{B} \cap \Gamma(X)$. Then we clearly have $g(x) \neq g(y)$.

3.4.11 Remark In uniformizable T_1 Cauchy spaces and more generally, in Cauchy subspaces of completely regular Hausdorff topological spaces, Γ(X) and C(X) are point separating and determine the same initial topology. However, without the complete regularity many examples of highly non-point-separating classes Γ(X) exist. For instance, if we take Hewitt's classical example of a regular topology (X,q) on which all real-valued continuous maps are constant, then any Cauchy structure on X compatible with q has a class Γ(X) which reduces to all constant maps.

When the conditions of the preceding proposition are fulfilled, that is, when $C(X)$ is point separating and $tC(X) = t\Gamma(X)$, then the fact that $\Gamma(X)$ is closed in $C(X)$ for pointwise or continuous convergence reduces to the equality of $\Gamma(X)$ and $C(X)$. This situation is studied in Chapter 4. We can already conclude that the Cauchy structure $\underline{C}_{G_c}$ on $\Gamma(X)$ is not necessarily complete.

In general (without the assumption $tC(X) = t\Gamma(X)$) the point separability of $C(X)$ does not imply the point separability of $\Gamma(X)$. We illustrate this statement by means of an example.

3.4.12 Example Let b and c be two points not belonging to the plane R^2. Let a be the origin of the plane. X is the set consisting of the point b and all points $x(r,\Theta)$ of the plane R^2 with polar coordinates (r,Θ), $r \in \mathsf{R}^+$, $\Theta \in [-\pi/3, 4\pi/3]$. Points $x \neq a$, $x \neq b$ have their usual neighborhoods. For $\varepsilon > 0$ let

$$V_a^{\varepsilon} = \{x(r,\Theta) \in \mathsf{R}^2 \mid r \in [0,\varepsilon),\ \Theta \in (\frac{5\Pi}{6}, \frac{4\Pi}{3}]\}$$

The sets V_a^{ε}, $\varepsilon > 0$ form a basis for the neighborhoods of a. For $\varepsilon > 0$ let

$$V_b^{\varepsilon} = \{x(r,\Theta) \in \mathsf{R}^2 \mid r \in (0,\varepsilon),\ \Theta \in [-\frac{\Pi}{3}, \frac{\Pi}{6})\}$$

The sets $V_b^{\varepsilon} \cup \{b\}$, $\varepsilon > 0$, form a basis for the neighborhoods of b. Next we prove that the collection $C(X)$ is point separating. To separate the points a and b take the function

$f : X \to \mathsf{R}$ which maps

b to $\frac{\Pi}{6}$

$x(r, \Theta)$ to $\frac{\Pi}{6}$ if $r > 0$ and $\Theta \in [-\frac{\Pi}{3}, \frac{\Pi}{6}]$

$x(r, \Theta)$ to Θ if $r > 0$ and $\Theta \in (\frac{\Pi}{6}, \frac{5\Pi}{6})$

$x(r, \Theta)$ to $\frac{5\Pi}{6}$ if $r \geq 0$ and $\Theta \in [\frac{5\Pi}{6}, \frac{4\Pi}{3}]$

When x and y are two other different points of X and if both are different from b, then take the restriction of a separating continuous function f for the usual topology on the plane and put f(b) = f(a).

Finally if, for instance, y = b, then take the restriction of a function f, continuous for the usual topology on the plane and separating x and a, and then again put f(b) = f(a). Now let $Y = X \cup \{c\}$. The points of X have the same neighborhoodbase as before and a base for the neighborhoods of c is formed by the sets $V_c^\varepsilon \cup \{c\}$, $\varepsilon > 0$, where

$$V_c^\varepsilon = \{x(r, \Theta) \mid r \in (0, \varepsilon), \ \Theta \in (\frac{\Pi}{6}, \frac{5\Pi}{6})\}$$

Y is a topological extension of X and it induces a Cauchy structure on X. Next we show that for this Cauchy structure $\Gamma(X)$ does not separate the points a and b.

Let f be Cauchy continuous and let $\hat{f}$ be its continuous extension to Y. Since a and c do not have disjoint closed neighborhoods in Y, it follows that $\hat{f}(a) = \hat{f}(c)$. For the same reason we also have $\hat{f}(b) = \hat{f}(c)$. Consequently, the equality f(a) = f(b) holds for every Cauchy continuous function.

3.5 DOMAINS

Again we restrict ourselves to T_1 Cauchy spaces $(X, \underline{C})$ with a point-separating class $\Gamma(X)$, and we make this extra assumption throughout this section. For any function class $\Phi \subset \mathbf{R}^X$ let $c\Phi$ be the initial Cauchy structure on X determined by the source $(X \xrightarrow{f} \mathbf{R}_t)_{f \in \Phi}$. Explicitly we have

$$c\Phi = \{ \underline{F} \in F(X) \mid \text{stack } f(\underline{F}) \text{ converges for every } f \in \Phi \}$$

Further we denote by

$t\Phi$ the initial topology on X determined by the source $(X \xrightarrow{f} R_t)_{f \in \Phi}$

$u\Phi$ the initial uniformity on X determined by the source $(X \xrightarrow{f} R_u)_{f \in \Phi}$

The point separability of $\Gamma(X)$ assures that the structures $t\Gamma(X)$, $c\Gamma(X)$ and $u\Gamma(X)$ are all Hausdorff.

The following result is essentially Theorem 2, Section 3, Part 14 in [24].

3.5.1 Proposition The three structures $t\Gamma(X)$, $c\Gamma(X)$, and $u\Gamma(X)$ are pairwise compatible.

Next we introduce some terminology due to Császár [31].

3.5.2 Definition If $\Phi \subset R^X$, then

i. A filter $\underline{F}$ on X is a Φ-filter if stack $f(\underline{F})$ converges for every f in Φ.

ii. Φ is filter closed if $f \in \Phi$ whenever stack $f(\underline{F})$ converges for every Φ-filter $\underline{F}$ on X.

So, using the terminology of Császár, $c\Gamma(X)$ is the collection of $\Gamma(X)$-filters.

3.5.3 Proposition The Cauchy structure $c\Gamma(X)$ has the following properties:

i. $c\Gamma(X,\underline{C})$ is coarser than $\underline{C}$.

ii. For a realvalued function f the following properties are equivalent:

1. $f:(X,u\Gamma(X)) \to R_u$ is uniformly continuous.
2. $f:(X,c\Gamma(X)) \to R_t$ is Cauchy continuous.
3. $f:(X,\underline{C}) \to R_t$ is Cauchy continuous.

Proof: (i) This follows by straightforward verification.

(ii) (1) $\Rightarrow$ (2) follows from the compatibility of $u\Gamma(X)$ and $c\Gamma(X)$. (2) $\Rightarrow$ (3) follows from (i) of this Proposition. (3) $\Rightarrow$ (1) follows from the very definition of $u\Gamma(X)$.

3.5.4 Remark The equivalence of ii(2) and ii(3) in the preceding Proposition is expressed by the following equality:

$$\Gamma(X,\underline{C}) = \Gamma(X,c\Gamma(X))$$

The nontrivial inclusion $\Gamma(X,c\Gamma(X)) \subset \Gamma(X,\underline{C})$ means that the function class $\Gamma(X,\underline{C})$ is filter closed in the sense of Definition 3.5.2. In [31] Császár has shown that a function class is filter closed if and only if it is composition closed. So using this equivalence the inclusion $\Gamma(X,c\Gamma(X)) \subset \Gamma(X,\underline{C})$ also follows from Proposition 3.2.4.

3.5.5 Corollary For two Cauchy structures $\underline{C}$ and $\underline{D}$ on X we have $\Gamma(X,\underline{C}) = \Gamma(X,\underline{D})$ if and only if $c\Gamma(X,\underline{C}) = c\Gamma(X,\underline{D})$.

Following the terminology in [8] and [96] we now introduce the following notion.

3.5.6 Definition A Cauchy space $(X,\underline{C})$ is a <u>domain</u> if

1. $\Gamma(X)$ separates the points of X.
2. $\underline{C}$ is the initial structure of $\Gamma(X)$.

3.5.7 Proposition A Cauchy space $(X,\underline{C})$ is a domain if and only if $\underline{C}$ is the initial structure of some point-separating subcollection of $\Gamma(X)$.

Proof: If $\Phi \subset \Gamma(X)$ is point separating and $c\Phi = \underline{C}$, then we clearly have

$$\underline{C} = c\Phi \leq c\Gamma(X) \leq \underline{C}$$

In view of Proposition 3.5.1 we immediately have the following result.

3.5.8 Proposition Every domain is uniformizable.

3.5.9 Proposition When $\Gamma(X)$ is point separating, $c\Gamma(X)$ is a domain. Moreover, the full subcategory of Chy whose objects are the domains is bireflective in the full subcategory of Chy whose objects are the Cauchy spaces with a point-separating class of real-valued Cauchy continuous maps. The bireflection of $(X,\underline{C})$ is

$$1_X : (X,\underline{C}) \to (X,c\Gamma(X))$$

Proof: From the equality $\Gamma(X,\underline{C}) = \Gamma(X,c\Gamma(X))$ it is clear that $c\Gamma(X)$ is a domain. Moreover, let $f:(X,\underline{C}) \to (Y,\underline{D})$ be Cauchy continuous, where $(Y,\underline{D})$ is a domain. For a filter $\underline{F} \in c\Gamma(X)$ and a function $g \in \Gamma(Y)$ we now have $g \circ f \in \Gamma(X)$, and then stack $g \circ f$ $(\underline{F})$ is convergent on the real line. By the arbitrariness of $g \in \Gamma(Y)$ we now can conclude that stack $f(\underline{F}) \in \underline{D}$ and finally, by the arbitrariness of $\underline{F} \in c\Gamma(X)$, we have that $f: (X,c\Gamma(X)) \to (Y,\underline{D})$ is Cauchy continuous.

3.5.10 Proposition When X has nonmeasurable cardinality then the following properties hold:

i. $(X,\underline{C})$ is a domain if and only if $(X,\underline{C})$ is uniformizable.

ii. When $\Gamma(X)$ is point separating, $1_X:(X,\underline{C}) \to (X,c\Gamma(X))$ is the bireflection of $(X,\underline{C})$ in the category of uniformizable Cauchy spaces.

Proof: (i) The "only if" part is Proposition 3.5.8. To prove the "if" part, suppose $(X,\underline{C})$ is uniformizable and $\underline{U}$ is a uniformity on X such that $\underline{C}_{\underline{U}} = \underline{C}$. Let $(\hat{X},\hat{\underline{U}})$ be the uniform completion of $(X,\underline{U})$ and $j: (X,\underline{U}) \to (\hat{X},\hat{\underline{U}})$ the embedding of $(X,\underline{U})$ into its completion. For $\underline{F} \in c\Gamma(X)$ and for every continuous function $h: (\hat{X},\hat{\underline{U}}) \to \mathbf{R}$ we have $h \circ j \in \Gamma(X)$ and hence stack $h \circ j(\underline{F})$ is convergent. It follows that stack $j(\underline{F})$ is a Cauchy filter on $(\hat{X},uC(\hat{X}))$. Since the cardinality of $\hat{X}$ is nonmeasurable, $uC(\hat{X})$ is a complete uniformity. It follows that stack $j(\underline{F})$ converges in the topology of $(\hat{X},\hat{\underline{U}})$ and finally we can conclude that $\underline{F} \in \underline{C}$.

(ii) This follows at once from Proposition 3.5.9 and (i) of this Proposition.

In the formulation of the next result we make the identification of Hausdorff convergence spaces with complete T_1 Cauchy spaces.

3.5.11 Proposition A Cauchy space is a complete domain if and only if it is a realcompact topological space.

Proof: Suppose $(X,\underline{C})$ is a complete domain. Then we can identify $\underline{C}$ with a Hausdorff convergence structure denoted by q. We then have

$$C(X,q) = \Gamma(X,\underline{C})$$

It follows that

$$c\mathrm{C}(X,q) = c\Gamma(X,\underline{C}) = \underline{C}$$

and

$$t\mathrm{C}(X,q) = q$$

So we can conclude that $u\mathrm{C}(X,q)$ is a compatible and complete uniformity for (X,q). Hence (X,q) is a realcompact topological space.

To prove the reverse implication, suppose (X,q) is a realcompact Hausdorff topological space. We identify q with the complete Cauchy structure, denoted by $\underline{C}$. Then we have

$$\Gamma(X,\underline{C}) = \mathrm{C}(X,q)$$

So clearly $\Gamma(X,\underline{C})$ is point separating, and moreover,

$$c\Gamma(X,\underline{C}) = c\mathrm{C}(X,q)$$

Since $c\mathrm{C}(X,q)$ is complete and compatible with q, it follows that

$$c\Gamma(X,\underline{C}) = \underline{C}$$

Finally we can conclude that $\underline{C}$ is a domain.

3.6 TOTALLY BOUNDED DOMAINS

As in Section 3.5 we make the assumption that $\Gamma(X)$ separates the points of X. Since $\Gamma(X)$ is an algebra, contains the constants, and is bounded inversion closed, we can apply Theorem 1.4 of Isbell's paper [61], which guarantees that under these conditions the collection $\Gamma^*(X)$ of bounded functions in $\Gamma(X)$ also separates the points of X and determines the same initial topology as $\Gamma(X)$.

Let $c\Gamma^*(X)$ be the initial Cauchy structure on X determined by the source $(X \xrightarrow{f} R_t)_{f \in \Gamma^*(X)}$. As in Section 3.5, $t\Gamma^*(X)$ denotes the initial topology of the source $(X \xrightarrow{f} R_t)_{f \in \Gamma^*(X)}$ and $u\Gamma^*(X)$ denotes the initial uniformity of the source $(X \xrightarrow{f} R_u)_{f \in \Gamma^*(X)}$.

3.6.1 Proposition The three structures $t\Gamma^*(X)$, $c\Gamma^*(X)$, and $u\Gamma^*(X)$ are pairwise compatible.

3.6.2 Proposition $c\Gamma^*(X)$ is totally bounded.

Proof: This follows from the fact that a bounded real-valued function maps an ultrafilter to a convergent filter.

3.6.3 Proposition $c\Gamma^*(X) = c\Gamma(X, c\Gamma(X) \cup U(X))$.

Proof: By Proposition 3.5.3 we have

$$\Gamma^*(X,\underline{C}) = \Gamma^*(X, c\Gamma(X))$$

Applying Proposition 3.3.5 we have

$$\Gamma^*(X,\underline{C}) = \Gamma(X, (c\Gamma(X))_t) = \Gamma(X, c\Gamma(X) \cup U(X))$$

Taking the initial Cauchy structures on both sides, we are done.

In view of Proposition 3.3.5, and by applying Propositions 3.5.3 and 3.5.5 to $\underline{C}_t$, we immediately obtain the following results.

3.6.4 Proposition The Cauchy structure $c\Gamma^*(X,\underline{C})$ has the following properties:

i. $c\Gamma^*(X,\underline{C})$ is coarser than $\underline{C}_t$

ii. For a real-valued function f the following properties are equivalent

1. f: $(X,u\Gamma^*(X)) \to \mathsf{R}_u$ is uniformly continuous.
2. f: $(X,c\Gamma^*(X)) \to \mathsf{R}_t$ is Cauchy continuous.
3. f: $(X,\underline{C}) \to \mathsf{R}_t$ is bounded and Cauchy continuous.

iii. For two Cauchy structures $\underline{C}$ and $\underline{D}$ on X we have $\Gamma^*(X,\underline{C}) = \Gamma^*(X,\underline{D})$ if and only if $c\Gamma^*(X,\underline{C}) = c\Gamma^*(X,\underline{D})$.

3.6.5 Proposition When $\Gamma(X)$ is point separating, $c\Gamma^*(X)$ is a totally bounded domain. Moreover, the full subcategory of Chy whose objects are the totally bounded domains is bireflective in the full subcategory of Chy whose objects are the Cauchy spaces with point-separating class of real-valued Cauchy continuous maps. The bireflection of $(X,\underline{C})$ is given by

$$1_X : (X,\underline{C}) \to (X,c\Gamma^*(X))$$

Proof: That $c\Gamma^*(X)$ is a totally bounded domain follows at once from Propositions 3.5.9 and 3.6.2. Moreover, 1_X: $(X,\underline{C}) \to (X,c\Gamma^*(X))$ is the composition of 1_X: $(X,\underline{C}) \to (X,\underline{C}_t)$, the reflection of $(X,\underline{C})$ in the category of all totally bounded Cauchy spaces, with 1_X: $(X,\underline{C}_t) \to (X,c\Gamma(X,\underline{C}_t))$, the reflection of $(X,\underline{C}_t)$ in the category of domains.

3.6.6 Proposition When X has nonmeasurable cardinality then the following properties hold:

i. $c\Gamma^*(X) = \underline{C}$ if and only if $\underline{C}$ is uniformizable and totally bounded.

ii. When $\Gamma(X)$ is point separating, 1_X: $(X,\underline{C}) \to (X,c\Gamma^*(X))$ is the bireflection of $(X,\underline{C})$ in the category of totally bounded uniformizable Cauchy spaces.

Proof: (i) If $\underline{C}$ is uniformizable and totally bounded then we have that $\underline{C} = \underline{C}_t$ is a domain. So $c\Gamma^*(X) = c\Gamma(X,\underline{C}_t) = \underline{C}_t = \underline{C}$. The reverse implication follows from Propositions 3.6.1 and 3.6.2. (ii) This follows at once from Proposition 3.6.5 and (i) of this Proposition.

As in Proposition 3.5.11 we identify Hausdorff convergence spaces with complete T_1 Cauchy spaces and obtain the analogous result for the totally bounded case.

3.6.7 Proposition A Cauchy space is a complete totally bounded domain if and only if it is a compact Hausdorff topological space.

Proof: Suppose $(X,\underline{C})$ is a complete totally bounded domain. Then by Proposition 3.5.11 $(X,\underline{C})$ can be identified with a Hausdorff realcompact topological space (X,q). Moreover, the totally boundedness of $(X,\underline{C})$ implies that (X,q) is compact.

To prove the reverse implication, suppose (X,q) is a compact topological Hausdorff space; then identifying q with the complete Cauchy structure denoted by $\underline{C}$ and again applying Proposition 3.5.11, we have that $(X,\underline{C})$ is a domain. The compactness of q moreover implies that $(X,\underline{C})$ is totally bounded.

4

A Comparison of the Classes of Continuous and the Classes of Cauchy Continuous Maps

As we already pointed out, every collection $C(X)$ of continuous maps is a collection $\Gamma(X)$, for a suitable Cauchy structure on X. In this chapter we answer the reverse question: When is a collection $\Gamma(X)$ of Cauchy continuous maps a collection $C(X)$ for a suitable convergence structure on X?

In Section 4.1 we make a comparison of the properties of collections $\Gamma(X)$ and collections $C(X)$, and we will be interested in the questions: What special properties do the $C(X)$'s have? How can we characterize the collections $C(X)$ among the classes $\Gamma(X)$?

The following problem is related to the questions treated in Section 4.1: Given a convergence structure q on X and a compatible Cauchy structure $\underline{C}$, when is $\Gamma(X,\underline{C}) = C(X,q)$? In Section 4.2 we investigate the effect of realcompactness of q and of completeness of $\underline{C}$.

In Section 4.3 the structure space theory for $\Gamma(X)$ is developed. It is shown that to a certain extent, this theory is similar to the theory of βX and υX built from $C(X)$. We also illustrate points of divergence of the two theories. These are mainly due to the fact that, in general, $\Gamma(X)$ is not inversion closed, and not even strongly bounded inversion closed, as was pointed out in (3.2.11).

Throughout this chapter we make the assumption that the Cauchy spaces involved are T_1, the convergence spaces are Hausdorff, and the function classes $\Gamma(X)$ or $C(X)$ all separate the points of X.

4.1 CHARACTERIZATION OF THE CLASSES C(X) AMONG THE CLASSES Γ(X)

First we reduce the question, "When is $\Gamma(X)$ equal to some collection $C(X)$?," to the following particular equality.

4.1.1 Lemma $\Gamma(X,\underline{C})$ is a collection $C(X,q)$ for a suitable convergence structure q if and only if $\Gamma(X,\underline{C}) = C(X,t\Gamma(X))$.

Proof: If $\Gamma(X,\underline{C}) = C(X,q)$ for some q, then $t\Gamma(X,\underline{C}) = tC(X,q) \leq q$ and consequently:

$$\Gamma(X,\underline{C}) \subset C(X,t\Gamma(X)) \subset C(X,q) = \Gamma(X,\underline{C})$$

The converse is evident.

Let $(X,\underline{C})$ be a Cauchy space and let $q_{\underline{C}}$ be the compatible convergence structure on X. Showing the equality of $\Gamma(X,\underline{C})$ and $C(X,q_{\underline{C}})$ can also be reduced to showing the equality of $\Gamma(X,\underline{C})$ and $C(X,t\Gamma(X))$.

4.1.2 Proposition If $\underline{C}$ and $q_{\underline{C}}$ are compatible, then the following properties are equivalent:

1. $\Gamma(X,\underline{C}) = C(X,q_{\underline{C}})$.
2. (i) $\Gamma(X,\underline{C}) = C(X,t\Gamma(X))$.
   (ii) $\Gamma(X,\underline{C})$ and $C(X,q_{\underline{C}})$ determine the same initial topology.
3. $c\Gamma(X,\underline{C}) = cC(X,q_{\underline{C}})$.

Proof: (1) $\Rightarrow$ (2) This follows immediately from Lemma (4.1.1).
(2) $\Rightarrow$ (3) Combining (i) and (ii) we have

$$\Gamma(X,\underline{C}) = C(X,tC(X,q_{\underline{C}})) = C(X,q_{\underline{C}})$$

Then by taking the initial Cauchy structures we obtain (3).

(3) $\Rightarrow$ (1) This follows at once from Corollary 3.5.5 applied to $\underline{C}$ and to $\underline{C}^{q_{\underline{C}}}$.

In the special case where $\underline{C}$ is a completely regular Cauchy subspace of a topological space, condition (ii) in (2) is superfluous. In this case we have $t\Gamma(X,\underline{C}) = q_{\underline{C}}$.

When we apply Proposition (4.1.2) to a domain, we obtain the following:

4.1.3 Corollary If $(X,\underline{C})$ is a domain and $q_{\underline{C}}$ is the compatible convergence structure, then the following are equivalent:

1. $\Gamma(X,\underline{C}) = C(X,q_{\underline{C}})$
2. Any $\underline{F}$ is a Cauchy filter if and only if every continuous function maps $\underline{F}$ to a convergent filter.

From the preceding chapter we recall that $\Gamma(X)$ is a uniformly closed algebra and a lattice containing the constants. Moreover $\Gamma(X)$ is composition closed (3.2.3), filterclosed (3.5.2), and bounded inversion closed (3.2.9). Moreover, by assumption, $\Gamma(X)$ separates the points of X. Of course, C(X), too, has all these properties, but it also satisfies some stronger ones, which generally do not hold for arbitrary collections $\Gamma(X)$. We already came across such properties in the preceding chapter in (3.2.2), (3.2.9), and (3.4.11). We shall now investigate whether these properties, strongly composition closedness, inversion closedness, and completeness for $\underline{C}_{G_c}$, are characterizing properties for the collections C(X) among the classes $\Gamma(X)$.

For strongly composition closedness the answer follows at once from the results of Császár [32]. Császár has shown that strongly composition closedness is necessary and sufficient for a function class Φ on X to be a collection $C(X,\tau)$ for some topology τ on X. From Lemma 4.1.1 and the result of Császár we immediately can make the following conclusion.

4.1.4 Proposition $\Gamma(X)$ is strongly composition closed if and only if it is equal to some collection C(X).

For completeness for the structure $\underline{C}_{G_c}$, the answer follows immediately from the results of Binz [21]. Let $(X,\underline{C})$ be an arbitrary Cauchy space and $q_{\underline{C}}$ its compatible convergence. Then $\Gamma(X,\underline{C})$ is a subalgebra of $C(X,q_{\underline{C}})$, containing the constants. If $\Gamma(X,\underline{C})$ is point separating then so is $\Gamma^*(X,\underline{C})$ and they both determine the same initial topology. If, moreover,

$$t\Gamma(X,\underline{C}) = tC(X,q_{\underline{C}})$$

then we can apply Binz's theorem [21], which we recalled in Proposition 3.4.9, and make the conclusion that $\Gamma(X,\underline{C})$ is dense in $C(X,q_{\underline{C}})$ for continuous convergence.

In view of Proposition 4.1.2 the following result is now immediately clear.

4.1.5 Proposition If $\Gamma(X,\underline{C})$ is point separating and determines the same initial topology as $C(X,q_{\underline{C}})$, then the following properties are equivalent:

1. $\Gamma(X,\underline{C})$ is complete for the Cauchy structure $\underline{C}_{\underline{G}_c}$ of continuous convergence.
2. $\Gamma(X,\underline{C})$ is equal to some collection C(X).

The third property we came across in Chapter 3 is inversion closedness. We showed that although every collection C(X) is inversion closed, $\Gamma(X)$ need not have this property. In contrast with strongly composition closedness and $\underline{C}_{\underline{G}_c}$-completeness, inversion closedness is not characterizing for the collections C(X) among the classes $\Gamma(X)$. We shall give some examples later on in (4.1.21) and in (4.1.22) of inversion closed collections $\Gamma(X)$ that are different from C(X) for every convergence structure on X.

In the next proposition, necessary and sufficient conditions are given for $\Gamma(X)$ to be inversion closed. The method of proof of this result is well known. We first need the following terminology:

4.1.6 Definition If Y is a convergence space and X is a subset of Y, then X is G_δ-<u>dense</u> in Y if every G_δ-set of Y intersects X.

4.1.7 Proposition $\Gamma(X)$ is inversion closed if and only if X is G_δ-dense in some strict completion of $(X, c\Gamma(X))$.

Proof: First, suppose $\Gamma(X)$ is inversion closed and let Y be the extension of X obtained by completing $u\Gamma(X)$. Suppose X is not G_δ-dense in Y. Let $G = \bigcap_{n>0} G_n$, where each G_n is open in Y and $G \cap X$ is empty, while G is not. Choose $a \in G$, and for each $n > 0$ a continuous function $0 \leq f_n \leq 1/2^n$ satisfying $f_n(a) = 0$ with f_n equal to $1/2^n$ on G_n^c. Then the function $h = \Sigma_{n>0} f_n$ is continuous on Y. By Proposition 3.5.3, $\Gamma(X,c\Gamma(X)) = \Gamma(X)$ and thus we have that the restriction f of h to X belongs to $\Gamma(X)$. Moreover, f is strictly positive on X. Since $h(a) = 0$ it follows that the function $1/f$ does not belong to $\Gamma(X)$.

Second, for the converse, suppose Y is a strict completion of $(X,c\Gamma(X))$ in which X is G_δ-dense. If f is a function in $\Gamma(X)$ which is never zero on X, then f has a continuous extension $\hat{f}$ over Y. Moreover, from the fact that

$$\hat{f}^{-1}(0) = \bigcap_{n>0} \{|\hat{f}| < n\}$$

it follows that $\hat{f}^{-1}(0)$ is a G_δ-subset of Y. Consequently, unless it is empty, $\hat{f}^{-1}(0)$ must intersect X.

Since inversion closedness alone is not strong enough to characterize the collections C(X) among the classes $\Gamma(X)$, we shall now introduce properties, always holding for C(X), but not always for $\Gamma(X)$. We investigate whether if, in combination with inversion closedness or with each other, they characterize the classes C(X).

4.1.8 Definition

1. If $\Phi \subset \mathsf{R}^X$ is a function class then a sequence $(x_n)_n$ in X is a Φ-sequence if $\langle x_n \rangle$ is a Φ-filter in the sense of Definition 3.5.2.
2. A function class $\Phi \subset \mathsf{R}^X$ is sequentially closed if $f \in \Phi$ whenever f is $t\Phi$-continuous and $(f(x_n))_n$ converges for every Φ-sequence $(x_n)_n$.

Sequential closedness clearly is a strengthening of filterclosedness. Moreover, every collection C(X) is sequentially closed, but in general $\Gamma(X)$ need not be sequentially closed. However, sequential closedness is not a characterizing property for the classes C(X). These assertions will become clear from the examples in (4.1.20), (4.1.21), (4.1.22) and (4.1.23).

In the next propositions we provide sufficient conditions for $\Gamma(X)$ to be sequentially closed.

4.1.9 Proposition If every nonconvergent Cauchy filter in $c\Gamma(X)$ contains a Cauchy filter with a countable base, then $\Gamma(X)$ is sequentially closed.

Proof: Suppose f is $t\Gamma(X)$ continuous and $(f(x_n))_n$ converges for every $\Gamma(X)$-sequence $(x_n)_n$. Let $\underline{F} \in c\Gamma(X)$.

Either $\underline{F}$ is $t\Gamma(X)$ convergent, and then $f(\underline{F})$ converges, or $\underline{F}$ contains a filter $\underline{G} \in c\Gamma(X)$ having a countable base $\{G_n \mid n \in \mathbb{N}\}$. For every sequence $(x_n)_n$ with $\langle x_n \rangle$ finer than $\underline{G}$, the sequence $(f(x_n))_n$ converges. Furthermore, it is clear that all such sequences have to converge to the same limit. Since

$$\underline{G} = \cap \{\langle x_n \rangle \mid \langle x_n \rangle \text{ finer than } \underline{G}\}$$

it follows that $f(\underline{G})$ converges and a fortiori, that $f(\underline{F})$ converges as well. Finally we can conclude that $f \in \Gamma(X, c\Gamma(X))$ and by the filterclosedness of $\Gamma(X)$ we thus have $f \in \Gamma(X)$.

In view of (3.5.10), we now have the following result:

4.1.10 Corollary Assuming X has nonmeasurable cardinality, every uniform structure on X for which every equivalence class of nonconvergent Cauchy filters has a minimum with a countable base, has a sequentially closed collection of Cauchy continuous maps. In particular, when X has nonmeasurable cardinality every metric structure on X has a sequentially closed collection of Cauchy continuous maps.

Finally, we need some conditions on zero sets and on their separation. We use the terminology of Mrówka [94] and Bentley and Taylor [18].

4.1.11 Definition Two nonempty disjoint sets A and B are separated by $\Phi \subset R^X$ if a, b $\in$ R implies there exists a function h $\in \Phi$ such that $h(A) \subset \{a\}$ and $h(B) \subset \{b\}$.

4.1.12 Definition A function class $\Phi \subset R^X$

1. Has the same zero sets as $C(X,t\Phi)$ if every zero set of a function in $C(X,t\Phi)$ is a zero set of a function in Φ.
2. Is discriminating if every pair of disjoint zero sets A,B of functions in Φ can be separated by Φ.
3. Separates the zero sets of $C(X,t\Phi)$ if every pair of disjoint zero sets A,B of functions in $C(X,t\Phi)$ can be separated by Φ.

Every function class C(X) has each of the three properties. Again, a general class $\Gamma(X)$ need not have these properties. Moreover, none of these properties is characterizing for the classes C(X). These assertions will become clear from the examples below. First we provide some sufficient and/or necessary conditions for $\Gamma(X)$ to have the properties defined above.

4.1.13 Definition A subspace X of Y is said to be z-embedded in Y if whenever Z is a zero set of C(X), there is a zero set Z' of C(Y) such that $Z' \cap X = Z$.

4.1.14 Proposition $\Gamma(X)$ has the same zero sets as $C(X,t\Gamma(X))$ if and only if X is z-embedded in some strict completion of $(X,c\Gamma(X))$.

Proof: Suppose $\Gamma(X)$ has the same zero sets as $C(X,t\Gamma(X))$ and let Y be any strict completion of $(X,c\Gamma(X))$. For a zero set Z of $C(X,t\Gamma(X))$ we then have:

$$Z = f^{-1}(0) \qquad \text{for some } f \in \Gamma(X)$$

which implies

$$Z = h^{-1}(0) \cap X \qquad \text{for the continuous extension h of f}$$

To prove the reverse implication, suppose X is z-embedded in the strict completion Y of $(X,c\Gamma(X))$. Let Z be a zero set of $C(X,t\Gamma(X))$. Then we have

$$Z = h^{-1}(0) \cap X \qquad \text{for some } h \in C(Y)$$

and for the restriction f of h to X, this implies

$$Z = f^{-1}(0)$$

Moreover, since h is continuous on Y the function f belongs to $\Gamma(X,c\Gamma(X))$. In view of the filterclosedness of $\Gamma(X)$ we can conclude that $f \in \Gamma(X)$.

It is well known that when $(X,t\Gamma(X))$ is either Lindelöf or almost compact, then every embedding of X is a z-embedding. So we immediately have the following corollary:

4.1.15 Corollary If $(X,t\Gamma(X))$ is Lindelöf or almost compact, then $\Gamma(X)$ has the same zero sets as $C(X,t\Gamma(X))$.

4.1.16 Proposition $\Gamma(X)$ is discriminating if and only if $\Gamma^*(X)$ is strongly bounded inversion closed (in the sense of Definition 3.2.9).

Proof: Suppose $\Gamma^*(X)$ is strongly bounded inversion closed. Let $A = f^{-1}(0)$, $B = g^{-1}(0)$ be disjoint, with f and g in $\Gamma(X)$, and let $a,b \in \mathbf{R}$ be arbitrary. Without loss of generality, we may suppose that f (and g) is bounded since $|f| \wedge 1$ (and $|g| \wedge 1$) also belongs to $\Gamma(X)$ and determines the same zero set as f (as g). Since $\Gamma^*(X)$ is strongly bounded inversion closed the function $f^2/(f^2 + g^2)$ belongs to $\Gamma^*(X)$. Let $k : \mathbf{R} \to \mathbf{R} : x \to k(x) = a + (b - a)x$. Then the function $h = k \circ f^2/(f^2 + g^2)$ belongs to $\Gamma^*(X)$ and fulfils the conditions $h(A) \subset \{a\}$ and $h(B) \subset \{b\}$.

The proof of the reverse implication will be postponed to Section 4.3 since it is based on results about the structure space of $\Gamma^*(X)$.

4.1.17 Corollary If $\Gamma(X)$ is inversion closed then it is discriminating.

4.1.18 Proposition $\Gamma(X)$ separates the zero sets of $C(X,t\Gamma(X))$ if and only if $\Gamma^*(X) = C^*(X,t\Gamma(X))$.

Proof: In view of Proposition 3.3.6, $\Gamma^*(X)$ is a subalgebra of $C^*(X,t\Gamma(X))$ containing the constants. In view of Proposition 3.3.5 and Corollary 3.4.2 it is, moreover, uniformly closed. So from the Hewitt approximation theorem [60] we have the following equivalence:

$\Gamma^*(X) = C^*(X,t\Gamma(X)) \Leftrightarrow$ For every pair A,B of disjoint zero sets of $C(X,t\Gamma(X))$ there is a function $f \in \Gamma^*(X)$ such that $\mathrm{cl}_{\mathbf{R}}\, f(A) \cap \mathrm{cl}_{\mathbf{R}}\, f(B) = \emptyset$. (1)

So it remains to be shown that $\Gamma(X)$ separates the zero sets in the sense of (4.1.12) if and only if $\Gamma^*(X)$ satisfies (1).

When A and B are disjoint, $a,b \in R$, $a < b$, and $f(A) \subset \{a\}$ and $f(B) \subset \{b\}$, then $(f \vee a) \wedge b$ has the same properties. So if $\Gamma(X)$ separates the zero sets of $C(X,t\Gamma(X))$, then so does $\Gamma^*(X)$. Moreover, then clearly condition (1) is fulfilled. The other implication can be shown using an argument analogous to 2.4 in [94]. Suppose $\Gamma^*(X)$ satisfies (1) for disjoint zero sets A and B of $C(X,t\Gamma(X))$. Let $a,b \in R$ be arbitrary. The closed sets $cl_R f(A)$ and $cl_R f(B)$ are disjoint. Using the normality of the real line we can find a continuous function $k \in [a,b]^R$ separating them. In view of Proposition 3.2.4 the function $k \circ f$ belongs to $\Gamma(X)$. Moreover, $k \circ f\ (A) \subset \{a\}$ and $k \circ f\ (B) \subset \{b\}$.

4.1.19 Proposition $\Gamma(X)$ separates the zero sets of $C(X,t\Gamma(X))$ if and only if $\Gamma(X)$ is discriminating and has the same zero sets as $C(X,t\Gamma(X))$.

Proof: The "if" part is trivial.

To prove the "only if" part, suppose $\Gamma(X)$ separates the zero sets of $C(X,t\Gamma(X))$. Clearly, then $\Gamma(X)$ is discriminating. Next we prove that $\Gamma(X)$ and $C(X,t\Gamma(X))$ have the same zero sets. Let $f \in C(X,t\Gamma(X))$. For $n \geq 1$, the sets $Z(f) = f^{-1}(0)$ and $|f|^{-1}[1/n, +\infty)$ are disjoint zero sets. So there exists a function $\phi_n \in \Gamma(X)$, $0 \leq \phi_n \leq 1/2^n$, mapping $Z(f)$ to 0 and $|f|^{-1}[1/n, +\infty)$ to $1/2^n$. The function $g = \Sigma_{n\geq 1} \phi_n$ belongs to $\Gamma(X)$, and moreover, we have

$$x \in Z(g) \Leftrightarrow x \in \bigcap_{n\geq 1} Z(\phi_n)$$

$$\Leftrightarrow \forall\, n \geq 1, |f|(x) < \frac{1}{n}$$

$$\Leftrightarrow x \in Z(f)$$

Next we give four examples of Cauchy structures on the real line and investigate the properties of their classes of Cauchy continuous maps. From these examples it will become clear that:

1. There are no other implications than those indicated in the diagram below. (These implications were shown in Propositions 4.1.17 and 4.1.19.)

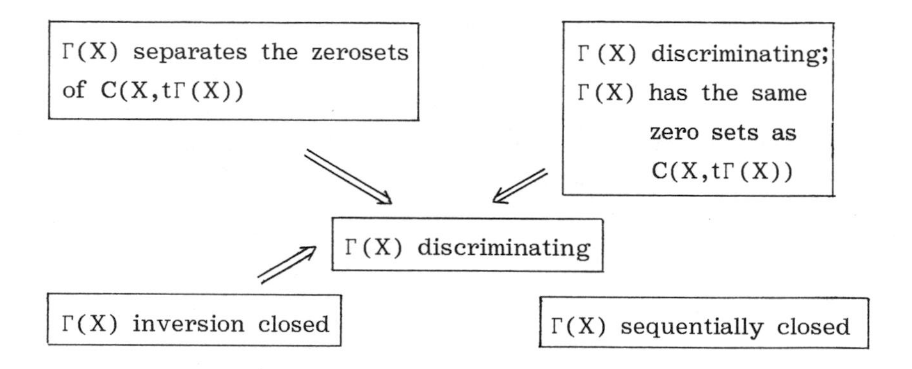

2. None of these properties alone, when assumed on Γ(X), implies that Γ(X) is a collection C(X). Even the combination of inversion closedness and sequential closedness of Γ(X) is not characterizing for C(X). The positive results are formulated in Theorems 4.1.24 and 4.1.26.

4.1.20 Example Let $X = \mathrm{R}\backslash\{0\}$ be equipped with the Cauchy structure induced by its extension R_t. Since this Cauchy structure is induced by a metric, Corollary 4.1.10 implies that Γ(X) is sequentially closed. Moreover, tΓ(X) is the usual topology on X, which is Lindelöf. So by Corollary 4.1.15, Γ(X) has the same zero sets as C(X).

The Cauchy continuous functions $x \vee 0$ and $(-x) \vee 0$ have disjoint zero sets that cannot be separated by a Cauchy continuous function. It follows that $\Gamma(X)$ is not discriminating, and hence neither separating the zero sets of C(X) nor inversion closed.

4.1.21 Example Let $X = \mathbb{R}$ and let $\underline{N}$ be the filter of finite complements on X. Take the uniformity $\underline{U} = [\Delta] \cap \underline{N} \times \underline{N}$ on X, where Δ is the diagonal of $X \times X$. Further, consider the Cauchy structure

$$\underline{C} = \{\dot{x} \mid x \in X\} \cup \{\underline{F} \in F(X) \mid \underline{F} > \underline{N}\}$$

This Cauchy structure is compatible with $\underline{U}$ and by Proposition 3.5.10 we have $c\Gamma(X) = \underline{C}$. Consequently $t\Gamma(X)$ is the discrete topology. We now show that the collection of Cauchy continuous maps

$$\Gamma(X) = \{f \in \mathbb{R}^{\mathbb{R}} \mid \text{stack } f(\underline{N}) \text{ converges}\}$$

is sequentially closed, inversion closed, and hence also discriminating.

To prove the first assertion we consider the point $p_{\underline{N}}$ in the completion $(\hat{X}, \hat{\underline{U}})$, corresponding to the minimal Cauchy filter $\underline{N}$. The singleton $\{p_{\underline{N}}\}$ is not a G_δ. Indeed, suppose on the contrary that there exists a countable collection $(G_n)_{n \in \mathbb{N}}$ of sets, open in $(\hat{X}, \hat{\underline{U}})$, such that $\{p_{\underline{N}}\} = \cap_{n \in \mathbb{N}} G_n$. Then for $n \in \mathbb{N}$ we could find $N_n \in \underline{N}$ with $N_n \subset G_n$. This would imply

$$\mathbb{R} = \{p_{\underline{N}}\}^c \subset \bigcup_{n \in \mathbb{N}} N_n^c$$

which is a contradiction. Consequently it follows that X is G_δ-dense in $(\hat{X}, \hat{\underline{U}})$, and applying Proposition 4.1.7 we are done.

To prove the second assertion, suppose f maps every $\Gamma(X)$-sequence to a convergent sequence on R. Every infinite sequence, for which all elements are different, is such a $\Gamma(X)$-sequence. Moreover the images by f of two such sequences have the same limits. So in view of

$$\underline{N} = \cap \{\langle x_n \rangle \mid (x_n)_n \text{ infinite sequence, } x_k \neq x_\ell \text{ if } k \neq \ell\}$$

we can conclude that stack $f(\underline{N})$ converges.

Next we show that $\Gamma(X)$ does not separate the zero sets of $C(X, t\Gamma(X))$. Take any partition of R into two infinite subsets A and B. The function 1_A, which is 1 on A and 0 on B, is a bounded function which is not Cauchy continuous. So in view of Proposition 4.1.18, $\Gamma(X)$ does not separate the zero sets. In view of Proposition 4.1.19 we also can conclude that $\Gamma(X)$ does not have the same zero sets as $(X, t\Gamma(X))$.

4.1.22 Example Again, let X = R, but now let $\underline{M}$ be the filter of countable complements on R and let $\underline{U} = [\Delta] \cap \underline{M} \times \underline{M}$. For $\underline{C}$ we take again the compatible Cauchy structure. We first show that, as in the previous example, the collection

$$\Gamma(X) = \{f \in R^R \mid \text{stack } f(\underline{M}) \text{ converges}\}$$

is inversion closed and hence also discriminating.

To prove this assertion, note that the point $\{p_{\underline{M}}\}$ corresponding to the minimal Cauchy filter $\underline{M}$ again is not a G_δ in the completion $(\hat{X}, \hat{\underline{U}})$. This follows from an argument similar to the one used in the previous example.

So again, X is G_δ-dense in the completion $(\hat{X}, \hat{\underline{U}})$ and $\Gamma(X)$ is inversion closed in view of Proposition 4.1.7. Next we prove that $\Gamma(X)$ is not sequentially closed. Let f be the identity on R. Since all $\Gamma(X)$-sequences are eventually constant, f maps every $\Gamma(X)$-sequence to a convergent sequence on R. However, f clearly is not Cauchy continuous.

Next we show that $\Gamma(X)$ does not separate the zero sets of $C(X,t\Gamma(X))$. To prove this, take any partition of R into two uncountable infinite sets A and B. The function 1_A is bounded but not Cauchy continuous. So in view of Proposition 4.1.18, $\Gamma(X)$ does not separate the zero sets. In view of Proposition 4.1.19 we also can conclude that $\Gamma(X)$ does not have the same zero sets as $C(X,t\Gamma(X))$.

4.1.23 Example Again, let X = R, but now let $\underline{U}$ be the uniformity of finite partitions and let $\underline{C}$ be the compatible Cauchy structure. Clearly $\underline{C} = U(R)$ and $t\Gamma(X)$ is the discrete topology. So we have $\Gamma(X) = \Gamma^*(X) = C^*(X,t\Gamma(X))$. Propositions 4.1.18 and 4.1.19 now imply that $\Gamma(X)$ separates the zero sets of $C(X,t\Gamma(X))$, $\Gamma(X)$ is discriminating, and has the same zero sets as $C(X,t\Gamma(X))$. On the other hand, however, the function e^{-x^2} is Cauchy continuous and strictly positive, whereas $1/e^{-x^2}$ is not Cauchy continuous. So $\Gamma(X)$ is not inversion closed. As in Example 4.1.22 all $\Gamma(X)$-sequences are eventually constant and thus the identity again maps every $\Gamma(X)$-sequence to a convergent sequence on R. Since it is, however, again not Cauchy continuous, this shows that $\Gamma(X)$ is not sequentially closed.

From the previous examples it follows that the only possible combinations of properties in the diagram above, characterizing the classes C(X) among the classes $\Gamma(X)$, are inversion closedness and separating the zero sets, or inversion closedness and same zero sets, or sequential closedness and separating the zero sets. We shall show that these combinations indeed give positive results.

4.1.24 Theorem The following properties are equivalent for any function class $\Gamma(X)$:

1. $\Gamma(X)$ is a collection $C(X)$ for a suitable convergence structure on X.
2. $\Gamma(X)$ is inversion closed and has the same zero sets as $C(X, t\Gamma(X))$.
3. $\Gamma(X)$ is inversion closed and separates the zero sets of $C(X, t\Gamma(X))$.

Proof: (1) $\Rightarrow$ (2) is trivial.

(2) $\Rightarrow$ (3) follows from Corollary 4.1.17 and Proposition 4.1.19.

(3) $\Rightarrow$ (1): In view of Proposition 3.2.1 and 3.4.2, the Mrówka version of the Stone-Weierstrass theorem [94] can be applied to the subalgebra $\Gamma(X)$ of $C(X, t\Gamma(X))$. This theorem implies that the two algebras are equal.

4.1.25 Remark As we pointed out earlier, $\Gamma(X)$ being inversion closed is not sufficient to imply the equality $\Gamma(X) = C(X, t\Gamma(X))$. However, in view of Corollary 4.1.15 and Theorem 4.1.24, inversion closedness is sufficient in case $(X, t\Gamma(X))$ is Lindelöf or almost compact. This fact is related to the results of Hager and Johnson [50] that a completely regular space X is Lindelöf or almost compact if and only if every uniformly closed, topology generating, inversion closed subalgebra of $C(X)$, containing the constants, equals $C(X)$. Examining the proof of this theorem, it becomes clear that the following also holds: A completely regular space X is Lindelöf or almost compact if and only if every inversion closed topology generating subcollection $\Gamma(X)$ of $C(X)$ equals $C(X)$.

Next we come to the second positive result.

4.1.26 Theorem $\Gamma(X)$ is a collection $C(X)$ for a suitable convergence structure on X if and only if $\Gamma(X)$ is sequentially closed and separates the zero sets of $C(X, t\Gamma(X))$.

Proof: We need only show the "if" part. Suppose $\Gamma(X)$ is sequentially closed and separates the zero sets of $C(X, t\Gamma(X))$. If $\Gamma(X) \neq C(X, t\Gamma(X))$, choose a function $f \in C(X, t\Gamma(X))$, $f \notin \Gamma(X)$. So we can find a $\Gamma(X)$-sequence $(x_n)_{n \in \mathbb{N}}$ such that $(f(x_n))_{n \in \mathbb{N}}$ does not converge. If $(f(x_n))_{n \in \mathbb{N}}$ is bounded, choose $M > 0$ such that $| f(x_n) | \leq M$ for every $n \in \mathbb{N}$. Let $h = f \wedge M \vee (-M)$. Then clearly we have $h \in C^*(X, t\Gamma(X))$ but $h \notin \Gamma(X)$.

In the other case, when $(f(x_n))_{n \in \mathbb{N}}$ is unbounded, for instance unbounded from above, choose a subsequence $(x_{k_n})_{n \in \mathbb{N}}$ such that $(f(x_{k_n}))_{n \in \mathbb{N}}$ is strictly increasing to $+\infty$. Now g is defined as follows: for $i \geq 0$ let

$$g(x) = \begin{cases} 1 & \text{for } x \leq f(x_{k_0}) \\ \dfrac{2(x - f(x_{k_{2i}}))}{f(x_{k_{2i}}) - f(x_{k_{2i+1}})} + 1 & \text{for } f(x_{k_{2i}}) \leq x \leq f(x_{k_{2i+1}}) \\ \dfrac{-2(x - f(x_{k_{2i+1}}))}{f(x_{k_{2i+1}}) - f(x_{k_{2i+2}})} - 1 & \text{for } f(x_{k_{2i+1}}) \leq x \leq f(x_{k_{2i+2}}) \end{cases}$$

Then $g \circ f \in C^*(X, t\Gamma(X))$, but $g \circ f \notin \Gamma(X)$ since $(g \circ f(x_{k_n}))_{n \in \mathbb{N}}$ does not converge. So in both cases a contradiction with Proposition 4.1.18 follows.

From Corollary 4.1.10 and the fact that on a metric space every closed set is a zero set, we immediately have the next result.

4.1.27 Corollary On a metric space of nonmeasurable cardinality, if the Cauchy continuous maps separate the closed sets, then every continuous function is Cauchy continuous.

As another consequence of Theorem 4.1.26, we obtain the following characterization of pseudocompactness.

4.1.28 Corollary A completely regular Hausdorff topological space (X, τ) is pseudocompact if and only if $C^*(X,\tau)$ is sequentially closed.

Proof: The only nontrivial part is sufficiency. Suppose $C^*(X,\tau)$ is sequentially closed. Let $\underline{C} = cC^*(X,\tau)$. Then we have $\Gamma(X,\underline{C}) = C^*(X,\tau)$ and $t\Gamma(X,\underline{C}) = \tau$. Consequently $\Gamma(X,\underline{C})$ separates the zero sets and by Theorem 4.1.26 we have $C^*(X,\tau) = \Gamma(X,\underline{C}) = C(X,t\Gamma(X)) = C(X,\tau)$.

4.2 THE EFFECT OF REALCOMPACTNESS AND COMPLETENESS

Let $(X,\underline{C})$ again be a Cauchy space and $q_{\underline{C}}$ be the compatible convergence structure. Whereas in the preceding section the equality of $\Gamma(X,\underline{C})$ and $C(X,q_{\underline{C}})$ was characterized by means of properties of $\Gamma(X,\underline{C})$ itself, in this section we shall study the effect of properties of $\underline{C}$ and $q_{\underline{C}}$. As we shall see in particular, completeness of $\underline{C}$ and realcompactness of $q_{\underline{C}}$ play an important role.

It is clear that, when $\underline{C}$ is a complete Cauchy structure and $q_{\underline{C}}$ is its compatible convergence, $\Gamma(X,\underline{C}) = C(X,q_{\underline{C}})$ holds. The next theorem treats the reverse implication.

4.2.1 Proposition A completely regular topological space (X,q) is realcompact if and only if the only compatible Cauchy structure $\underline{C}$ for which $\Gamma(X,\underline{C}) = C(X,q)$ is the complete one.

Proof: Suppose q is realcompact and let $\underline{C}$ be a compatible Cauchy structure for which $\Gamma(X,\underline{C}) = C(X,q)$. Then clearly we have

$$\underline{F} \in \underline{C} \Rightarrow \underline{F} \in c\Gamma(X,\underline{C})$$
$$\Rightarrow \underline{F} \in cC(X,q)$$

which, from the realcompactness of q, implies that $\underline{F}$ q-converges. For the converse, suppose q is completely regular but not realcompact. So $cC(X,q)$ is compatible but not complete. However, in view of the filterclosedness of $C(X,q)$ we have $\Gamma(X,cC(X,q)) = C(X,q)$.

4.3 STRUCTURE SPACES OF $\Gamma(X)$

Let $(X,\underline{C})$ be a Cauchy space. As we mentioned earlier, $\Gamma(X)$ and $\Gamma^*(X)$ determine the same initial topology on X which, in view of the point-separability assumption on $\Gamma(X)$, is a completely regular Hausdorff topology. We denote

$$t\Gamma(X) = t\Gamma^*(X) = \tau$$

As we already mentioned in Chapter 3, many extensions Y of (X, τ) (i.e., Hausdorff convergence spaces containing (X,τ) as a dense subspace) have a collection $C(Y)$ of continuous maps satisfying $C(Y)/X = \Gamma(X)$. For instance, every strict completion of $(X,c\Gamma(X))$ satisfies this condition. The same is true for $\Gamma^*(X)$. Every strict completion Y of $(X,c\Gamma^*(X))$ is an extension of (X,τ) satisfying $C(Y)/X = \Gamma^*(X)$. However, in this section, we prove that if we restrict ourselves to realcompact topological extensions (or if we restrict ourselves to compact topological extensions in the case of $\Gamma^*(X)$), then the condition $C(Y)/X = \Gamma(X)$ (or $C(Y)/X = \Gamma^*(X)$) determines a "unique" space Y. Here uniqueness has the usual meaning, that is, up to the following equivalence: Two extensions Y_1 and Y_2 are <u>equivalent</u> if and only if there is a homeomorphism between them leaving X pointwise fixed. This equivalence is denoted by

$$Y_1 \approx Y_2$$

We first investigate the case of the compact extension. We recall the constructions of a compact structure space developed in the literature. For more information the reader is referred to [48], [54], and [61].

4.3.1 Definition Let $\underline{A} \subset \mathsf{R}^X$ be a point-separating, uniformly closed function algebra and lattice containing the constants.

1. $\underline{H}(\underline{A}^*)$ is the set of real-valued homomorphisms on $\underline{A}^*$, topologized by the initial topology of the source

   $$\{\bar{a} \mid a \in \underline{A}^*\}$$

   where

   $$\bar{a} : \underline{H}(\underline{A}^*) \to \mathsf{R}$$
   $$h \to \bar{a}(h) = h(a)$$

2. $\hat{X}^*$ is the topological space obtained by completing the uniform space $(X, u\underline{A}^*)$ and then taking the topology of this completion.
3. $\underline{M}(\underline{A}^*)$ is the set of maximal ideals of $\underline{A}^*$, endowed with the Stone topology with base for the closed sets

   $$\{\underline{S}(a) \mid a \in \underline{A}^*\}$$

   where

$\underline{S}(a) = \{M \in \underline{M}(\underline{A}^*) \mid a \in M\}$

This topology is known to be equivalent to the hull-kernel topology defined by means of the following closure operator on subsets of $\underline{M}(\underline{A}^*)$:

$\text{cl}\,\underline{L} = \{M \in \underline{M}(\underline{A}^*) \mid \underline{M} \supset \cap\underline{L}\}$

The space $\underline{M}(\underline{A})$ of maximal ideals of $\underline{A}$ is defined analogously. Under the assumption made on $\underline{A}$, the spaces $\underline{M}(\underline{A})$ and $\underline{M}(\underline{A}^*)$ are equivalent.

4. For every $f \in \underline{A}^*$ the image $f(X)$ is contained in some compact interval $[a_f, b_f]$.

Let $E = \prod_{f \in \underline{A}^*} [a_f, b_f]$ and

$h : X \to E$

$x \to h(x) = (f(x))_{f \in \underline{A}^*}$

Then the set $\text{cl}_E h(X)$ is endowed with the topology induced by E.

The following result is well known.

4.3.2 Proposition If $\underline{A} \subset R^X$ is a point separating uniformly closed function algebra and a lattice containing the constants, then there is a unique topological compactification Y of $(X, t\underline{A})$ such that $C(Y)/X = \underline{A}^*$.

The unique space Y can be obtained by any one of the equivalent constructions mentioned in Definition 4.3.1: (1), (2), (3), and (4). We shall use Isbell's notation and put $Y = \underline{H}(\underline{A}^*)$.

In view of Propositions 3.2.1 and 3.4.2 we can apply Proposition 4.3.2 to $\Gamma(X)$. $\underline{H}(\Gamma^*(X))$ is the unique topological compactification of $(X, t\Gamma(X))$ satisfying $C(\underline{H}\Gamma^*(X))/X = \Gamma^*(X)$. It will be called the <u>compact structure space</u> of $\Gamma(X)$.

As a topological compactification of $(X, t\Gamma(X))$, satisfying $C(\underline{H}(\Gamma^*(X)) \supset X = \Gamma^*(X)$, the space $\underline{H}(\Gamma^*(X))$ is unique. But it is not unique among all compactifications of $(X, t\Gamma(X))$ with this property. We now illustrate this fact.

4.3.3 Example Let X be the real line endowed with the discrete topology, and let $\underline{U}_0$ be a fixed nonprincipal ultrafilter and $\underline{E}$ the filter of finite complements on X. We define a first compatible Cauchy structure $\underline{C}$ on X by decreeing that $\underline{F}$ is a Cauchy filter if and only if $\underline{F}$ is either a principal ultrafilter or $\underline{F}$ is finer than $\underline{E}$. Then clearly

$$\Gamma(X, \underline{C}) = \{ f \in R^X \mid \text{stack } f(\underline{E}) \text{ converges} \}$$

and $t\Gamma(X, \underline{C})$ is again the discrete topology on X. The compact structure space of $\Gamma(X, \underline{C})$ is equivalent to the Alexandroff compactification of X.

Next we consider a second Cauchy structure $\underline{D}$, compatible with X, be decreeing that $\underline{F}$ is a Cauchy filter if and only if $\underline{F}$ is either a principal ultrafilter or $\underline{F} = \underline{U}_0$ or $\underline{F} > \underline{E}$ and $\underline{U}_0 \not> \underline{F}$. We prove that $\Gamma(X, \underline{D}) = \Gamma(X, \underline{C})$.

Suppose $f \in \Gamma(X, \underline{D})$. Then stack $f(\underline{F})$ converges whenever $\underline{F} > \underline{E}$ and $\underline{U}_0 \not> \underline{F}$. Moreover, all these filters stack $f(\underline{F})$ have to converge to the same limit. In view of the equality

$$\text{stack } f(\underline{E}) = \text{stack } f(\cap\{ \underline{F} \mid \underline{F} > \underline{E}, \underline{U}_0 \not> \underline{F} \})$$

$$= \cap\{ \text{stack } f(\underline{F}) \mid \underline{F} > \underline{E}, \underline{U}_0 \not> \underline{F} \}$$

we now can conclude that stack $f(\underline{E})$ also converges. Hence $f \in \Gamma(X, \underline{C})$. The other inclusion follows immediately since $\underline{D} \geq \underline{C}$.

Now let Z be a completion of $(X,\underline{D})$ of the type $K_\omega(X,\underline{D})$, where ω maps on ultrafilters, as was described in Definition 1.4.10. It follows from the properties of this completion, developed in Proposition 1.4.11 and Theorem 2.4.4, that Z is a non-topological Hausdorff compactification of X satisfying $C(Z)/X = \Gamma(X,\underline{C})$. Clearly Z is not equivalent to the compact structure space of $\Gamma(X,\underline{C})$.

4.3.4 Proposition $\underline{H}(\Gamma^*(X))$ is a complete totally bounded domain. Moreover, the full subcategory of Chy, whose objects are the complete totally bounded domains, is reflective in the full subcategory of Chy whose objects are the Cauchy spaces with point separating class of Cauchy continuous maps. The reflection of $(X,\underline{C})$ is given by

$$j : (X,\underline{C}) \to \underline{H}(\Gamma^*(X))$$
$$x \to j(x)$$

where $j(x)(f) = f(x)$.

Proof: When we identify the topology of $\underline{H}(\Gamma^*(X))$ with the complete Cauchy structure, and apply Proposition 3.6.7 it then follows that $\underline{H}(\Gamma^*(X))$ is a complete totally bounded domain. Moreover, from Proposition 3.6.5 we recall that

$$1_X : (X,\underline{C}) \to (X,c\Gamma^*(X))$$

is the bireflection of $(X,\underline{C})$ in the category of totally bounded domains. By (2) in Definition 4.3.1, $\underline{H}(\Gamma^*(X))$ is equivalent to the strict uniformizable completion of $(X,c\Gamma^*(X))$ with embedding map i. Consequently

$$i : (X,c\Gamma^*(X)) \to \underline{H}(\Gamma^*(X))$$

is the reflection of the totally bounded domain $(X, c\Gamma^*(X))$ in the category of complete totally bounded domains. Finally, the reflection $j : (X, \underline{C}) \to \underline{H}(\Gamma^*(X))$ is obtained as a composition of i and 1_X.

Up to now the development of the theory for the compact structure space of $\Gamma(X)$ is quite similar to the theory of the Stone-Čech compactification βX developed from C(X). However, as done in [48], βX can also be obtained by using the Wallman method on the collection of zero sets of C(X).

We shall now investigate whether a similar construction can be applied to $\Gamma(X)$, and if so, whether it gives rise to a compactification equivalent to $\underline{H}(\Gamma^*(X))$. We first recall some notions and results from the literature about the generalized Wallman process for compactifying a space [5, 40, 105]. Frink [40] defined the concept of a normal base F on a Tychonoff space X in the following way.

4.3.5 Definition A base F for the closed subsets of a Tychonoff space is a <u>normal base</u> if and only if

1. F is a lattice of closed subsets, which means
   i. $\emptyset, X \in F$.
   ii. If $A, B \in F$, then $A \cap B$ and $A \cup B \in F$.
2. F is disjunctive, which means: if $A \in F$ and $x \in X \setminus A$ then there exists $B \in F$ with $x \in B$ and $A \cap B = \emptyset$.
3. F is normal, which means: if $A, B \in F$ are disjoint then there exist $C, D \in F$ with A and D disjoint, B and C disjoint, and $C \cup D = X$.

Frink applied Wallman's construction [112] to a normal base F and obtained a Hausdorff compactification w(F) of X in the following way: w(F) is the set of all F-ultrafilters. This set becomes a space by taking as a closed base the collection

$\{ A^* \mid A \in F \}$

where A^* is the set of all F-ultrafilters having A as a member.

For $\underline{A} \subset C(X)$ the first question is whether $Z(\underline{A})$, the zero sets of functions in $\underline{A}$, is a normal base. In relation to this question Biles [19] introduced the following notion and proved the next result.

4.3.6 Definition A subring $\underline{A}$ of $C(X)$ is a Wallman ring on a completely regular space X provided $Z(\underline{A})$, the zero sets of functions in $\underline{A}$, is a normal base on X.

4.3.7 Proposition Let X be a completely regular space and let $\underline{A}$ be a subring of $C(X)$ such that

1. $Z(\underline{A})$ is a base for the closed subsets of X.
2. $f \in \underline{A} \Rightarrow |f| \in \underline{A}$.

Then $\underline{A}$ is a Wallman ring on X.

The preceding result, when applied to the case of $\Gamma^*(X)$, will guarantee that we can make the Wallman construction of $Z(\Gamma^*(X))$. To compare $wZ(\Gamma^*(X))$ with $\underline{H}(\Gamma^*(X))$ we still need some definitions and results of Bentley and Taylor [18] and of Hager [49]. For the proof of Propositions 4.3.8 and 4.3.10 we refer to [18] and [49], respectively.

4.3.8 Proposition Let Y be a compactification of a completely regular space X, and $\underline{A}_Y$ the subset of $C(X)$ consisting of those real-valued continuous functions on X that are continuously extendible to all of Y. Then Y is equivalent to $w(Z(\underline{A}_Y))$ if and only if $\underline{A}_Y$ is discriminating (in the sense of Definition 4.1.12).

4.3.9 Definition Let Y be a compactification of a completely regular space X, and let $\underline{A}_Y$ be the subset of C(X) consisting of those real-valued continuous functions on X that are continuously extendible to all of Y. Then there is a <u>smallest inversion closed, topology generating, uniformly closed subalgebra of C(X), that contains the constants and is larger than $\underline{A}_Y$.</u> This algebra is denoted by $[\underline{A}_Y]$.

4.3.10 Proposition Let X be a completely regular topological space and let Y, $\underline{A}_Y$, and $[\underline{A}_Y]$ be defined as in Definition 4.3.9. Then the spaces $\underline{H}([\underline{A}_Y]^*)$ and $w(Z(\underline{A}_Y))$ are equivalent compactifications of X.

We now combine the previous results and apply them to $\Gamma^*(X)$.

4.3.11 Theorem Let $(X,\underline{C})$ be a Cauchy space and put $\tau = t\Gamma(X)$ as the initial topology. Then the following hold:

1. $Z(\Gamma(X)) = Z(\Gamma^*(X))$ is a normal base on (X,τ).
2. $wZ(\Gamma^*(X))$ is a Hausdorff compactification of (X,τ).
3. The following properties are equivalent:
   i. $wZ(\Gamma^*(X)) \approx \underline{H}(\Gamma^*(X))$.
   ii. $\Gamma^*(X)$ is strongly bounded inversion closed (3.2.9).
   iii. $\Gamma(X)$ is discriminating (4.1.12).
   iv. $\Gamma^*(X)$ is discriminating.

Proof: (1) That $\Gamma(X)$ and $\Gamma^*(X)$ determine the same zero sets follows at once, since we pointed out earlier that f, $|f|$, and $|f| \wedge 1$ all determine the same zero set and belong to $\Gamma(X)$ whenever $f \in \Gamma(X)$. Since $\Gamma(X)$ is an algebra and a lattice with constants, $Z(\Gamma(X))$ is a base for the closed sets of (X,τ). Proposition 4.3.7 now immediately implies that $Z(\Gamma(X))$ is a normal base.

2. Follows at once from (1) and the results of Frink.

3. (i) $\Rightarrow$ (ii) Suppose $wZ(\Gamma^*(X)) \approx \underline{H}(\Gamma^*(X))$. We apply Proposition 4.3.10 to (X,τ) and $Y = \underline{H}(\Gamma^*(X))$. Then $\underline{A}_Y$, the subset of $C(X)$ consisting of those functions which are continuously extendible to all of Y, is exactly $\Gamma^*(X)$. Let $[\underline{A}_Y]$ be defined as in Definition 4.3.9. It follows that the compactifications $\underline{H}(\Gamma^*(X))$ and $\underline{H}([\underline{A}_Y]^*)$ are equivalent. Now let $g,f \in \Gamma^*(X)$ be such that f is never zero on X and g/f is a bounded function. Then clearly g/f belongs to $[\underline{A}_Y]^*$. So g/f has a continuous extension to $\underline{H}([\underline{A}_Y]^*)$ and in view of the equivalence, also to $\underline{H}(\Gamma^*(X))$. So clearly g/f belongs to $\Gamma(X)$.

(ii) $\Rightarrow$ (iii) was shown in Proposition 4.1.16.

(iii) $\Rightarrow$ (iv) If A and B are disjoint zero sets of functions in $\Gamma^*(X)$ and $a < b \in \mathsf{R}$, then by (iii) there exists a function $h \in \Gamma(X)$, $h(A) \subset \{a\}$, $h(B) \subset \{b\}$. But then the function $h' = (h \vee a) \wedge b$ belongs to $\Gamma^*(X)$ and $h'(A) \subset \{a\}$, $h'(B) \subset \{b\}$.

(iv) $\Rightarrow$ (i) Suppose $\Gamma^*(X)$ is discriminating. We apply Proposition 4.3.8 to (X,τ) and $Y = \underline{H}(\Gamma^*(X))$. Then $\underline{A}_Y = \Gamma^*(X)$. So we immediately have $\underline{H}(\Gamma^*(X)) \approx wZ(\Gamma^*(X))$.

Next we turn to the realcompact extensions. We recall the constructions of a realcompact structure space developed in the literature.

4.3.12 Definition Let $\underline{A} \subset \mathsf{R}^X$ be a point separating, uniformly closed, function lattice and algebra, containing the constants:

1. $\underline{H}(\underline{A})$ is the set of real-valued homomorphisms on $\underline{A}$, with the initial topology of the source

   $\{\overline{a} \mid a \in \underline{A}\}$

   where $\overline{a} : \underline{H}(\underline{A}) \to \mathsf{R}$

   $h \to \overline{a}(h) = h(a)$

2. $\hat{X}$ is the topological space obtained by completing the uniform space $(X, u\underline{A})$ and then taking the topology of this completion.
3. $\underline{R}(\underline{A})$ is the subspace of $\underline{M}(\underline{A})$ consisting of the real ideals, that is,

$$\underline{R}(\underline{A}) = \{ M \in \underline{M}(\underline{A}) \mid \underline{A} \mid_M \approx \mathsf{R} \}$$

4. For every $f \in \underline{A}$ let R_f be a copy of the real line. Further let $\Pi_{f \in \underline{A}} \mathsf{R}_f$ be the product space and let

$$h : X \to \Pi \mathsf{R}_f$$
$$x \to h(x) = (f(x))_{f \in \underline{A}}$$

Then the set $\mathrm{cl}_{\Pi \mathsf{R}_f} h(X)$ is endowed with the topology induced by $\Pi_{f \in \underline{A}} \mathsf{R}_f$.

It is well known that, under the assumptions made on $\underline{A}$, the spaces constructed in Definition 4.3.12, (1), (2), (3), and (4) are all equivalent realcompactifications of $(X, t\underline{A})$. Moreover, for each of these extensions Y we have the inclusion

$$\underline{A} \subset C(Y)/X$$

However the assumptions made on $\underline{A}$ are not sufficient to have the equality $\underline{A} = C(Y)/X$.

The situation of Y being a realcompactification of X was treated in Section 1.4. Now, as we know, $C(Y)/X = \Gamma(X, \underline{C})$ for the Cauchy structure $\underline{C}$ induced by the extension Y. It follows from Proposition 3.2.4 that $\underline{A}$, in order to satisfy $\underline{A} = C(Y)/X$, needs to be composition closed. Császár has shown that this condition is also sufficient [31.

4.3.13 Proposition If X is a completely regular space and $\underline{A}$ is a subclass of C(X) such that

1. $Z(\underline{A})$ is a base for the closed sets of X
2. $\underline{A}$ is composition closed

then there is a "unique" topological realcompactification Y of X such that $C(Y)/X = \underline{A}$.

To prove the theorem, the space Y constructed by Császár is the one of Definition 4.3.12 (4). In fact, Y can be obtained by any of the constructions of Definition 4.3.12. We use Isbell's notation and put $Y = \underline{H}(\underline{A})$.

In view of Proposition 3.2.4, we can now apply the previous theorem to $\Gamma(X)$ and it then follows that $\underline{H}(\Gamma(X))$ is the unique topological realcompactification of $t(\Gamma(X))$, such that $C(\underline{H}(\Gamma(X))/X = \Gamma(X)$. $\underline{H}(\Gamma(X))$ will be called the realcompact structure space of $\Gamma(X)$.

4.3.14. Proposition $\underline{H}(\Gamma(X))$ is a complete domain. Moreover, the full subcategory of Chy, whose objects are the complete domains, is reflective in the full subcategory of Chy, whose objects are the Cauchy spaces with point separating class of Cauchy continuous maps. The reflection of $(X,\underline{C})$ is given by

$$j : (X,\underline{C}) \to \underline{H}(\Gamma(X))$$

where $j(x)(f) = f(x)$.

Proof: Completely analogous to the proof of Proposition 4.3.3.

Now if the construction of the realcompact structure space is applied to $C(X)$, it is well known that one obtains the Hewitt realcompactification υX. We also know that υX can be constructed by using the Wallman process and by describing υX as a subspace of $wZ(C(X))$. We now investigate whether a similar construction, when applied to $\Gamma(X)$, gives rise to $\underline{H}(\Gamma(X))$.

Let $rZ(\underline{A})$ be the subspace of $wZ(\underline{A})$ consisting of all $Z(\underline{A})$-ultrafilters with the countable intersection property. It is well known that, under the assumptions made for $\underline{A}$, $rZ(\underline{A})$ is a realcompactification of $t(\underline{A})$. We need the following result of Hager [49].

4.3.15 Proposition Let Y be a compactification of a completely regular space X, and let $\underline{A}_Y$ and $[\underline{A}_Y]$ be defined as in Definition 4.3.9. Then the extensions $\underline{H}([\underline{A}_Y])$ and $rZ(\underline{A}_Y)$ are equivalent.

4.3.16 Theorem The realcompact extensions $\underline{H}(\Gamma(X))$ and $rZ(\Gamma(X))$ of $t\Gamma(X)$ are equivalent if and only if $\Gamma(X)$ is inversion closed.

Proof: Let $Y = \underline{H}(\Gamma^*(X))$; then as before, $\underline{A}_Y = \Gamma^*(X)$. Applying Proposition 4.3.15, we obtain the equivalence of $\underline{H}([\underline{A}_Y])$ and $rZ(\Gamma(X))$. To prove the "if" part, suppose $\Gamma(X)$ is inversion closed. Since $\underline{A}_Y \subset \Gamma(X)$, it follows also that $[\underline{A}_Y] \subset \Gamma(X)$. To show the other inclusion, let f be any function in $\Gamma(X)$. By applying the inversion closedness of $\Gamma(X)$ and the definition of $[\underline{A}_Y]$, we have the following implications:

$$f \in \Gamma(X) \Rightarrow 1 + f^2 \in \Gamma(X)$$

$$\Rightarrow \frac{1}{1 + f^2} \in \Gamma^*(X)$$

$$1 + f^2 \in [\underline{A}_Y] \quad (1)$$

By applying the same argument on 1 + f, we have:

$$f \in \Gamma(X) \Rightarrow 1 + f \in \Gamma(X)$$

$$\Rightarrow (1 + f)^2 \in [\underline{A}_Y] \quad (2)$$

Combining (1) and (2) it follows that $f \in [\underline{A}_Y]$. Thus we have shown that $[\underline{A}_Y] = \Gamma(X)$ and we can conclude that $\underline{H}(\Gamma(X)) \approx rZ(\Gamma(X))$.

To prove the "only if" part, suppose $\underline{H}(\Gamma(X))$ and $rZ(\Gamma(X))$ are equivalent. Then in view of Proposition 4.3.15 $\underline{H}(\Gamma(X))$ and $\underline{H}([\underline{A}_Y])$ are equivalent extensions of $(X, t\Gamma(X))$. It is well known that X is G_δ-dense in $\underline{H}([\underline{A}_Y])$ and thus also in $\underline{H}(\Gamma(X))$. In view of Proposition 4.1.7, it follows that $\Gamma(X)$ is inversion closed.

4.3.17 Remark While the constructions of βX and υX, by means of ideals in $C(X)$, or by means of ultrafilters in $Z(C(X))$, are equivalent, Theorems 4.3.11 and 4.3.16 show that this is generally not the case for the constructions when applied to $\Gamma(X)$. The reasons are explained in Theorems 4.3.11 and 4.3.16. We can illustrate them somewhat further, by noting there is a certain lack of parallelism between ideals in $\Gamma(X)$ and ultrafilters in $Z(\Gamma(X))$. If I is a proper ideal in $\Gamma(X)$, then every $f \in I$ is noninvertible in $\Gamma(X)$. This however does not guarantee that $f^{-1}(0)$ is nonempty. So $\{f^{-1}(0) \mid f \in I\}$ is not necessarily a filterbase in the collection of zero sets. The parallelism can only be established between so-called "filter ideals" and ultrafilters in the collection of zero sets. Such a theory was developed by Biles for arbitrary Wallman rings [19]. In view of his theorem to which we referred earlier in Proposition 4.3.7, the theory of "filter ideals" can be applied to $\Gamma(X)$.

5

Characterizing the Function Classes $\Gamma(X)$

In this chapter we treat the following problem. Given a point separating function class Φ on a set X, find necessary and sufficient conditions on Φ for the existence of a Cauchy structure on X such that

$$\Phi = \Gamma(X)$$

A first solution will be established, using the properties of the classes $\Gamma(X)$ that we proved in Section 3.2. We shall show that composition closedness is one such characterizing property for the classes $\Gamma(X)$. A second solution to the problem will be based on two properties of Φ of a different type, a composition property (weaker than composition closedness) combined with a completeness condition.

For bounded function classes a solution has been known for years, and is a corollary of Proposition 4.3.2. A bounded point separating function class is a collection $\Gamma(X)$, for a suitable Cauchy structure, if and only if it is a uniformly complete function algebra containing the constants. However, this theorem does not extend to the unbounded case. In fact, uniform completeness is not strong enough. On the other hand, as we saw in Section 3.4, completeness for the structures of pointwise or continuous convergence are generally too strong, since classes $\Gamma(X)$ need not be complete for these structures. We shall introduce another completeness notion for a function class Φ by means of a natural Cauchy structure γ_Φ on Φ. This structure is related to the Cauchy continuous structure $\hat{\underline{C}}$ of Gazik and Kent, described in Section 1.3. In our main theorem we shall prove that a point separating function class Φ is a collection $\Gamma(X)$ for a suitable Cauchy structure on X if and only if Φ is a γ_Φ-complete function algebra containing the constants.

On bounded function classes Φ uniform completeness coincides with γ_Φ-completeness. So as a corollary, we obtain the result formulated above for bounded function classes. We also get another bonus from our main theorem. A combination of our first and second characterization can be used to formulate necessary and sufficient conditions for a function class to be composition closed, thus answering a problem posed by Császár in [32].

Throughout the chapter, Φ is supposed to be a point separating function class. Cauchy spaces satisfy the T_1 condition, and convergence spaces are supposed to be Hausdorff.

5.1 COMPOSITION CLOSED FUNCTION CLASSES

Let Φ be a function class on X and $c\Phi$ the initial Cauchy structure on X determined by Φ. In the terminology of A. Császár, $c\Phi$ is the collection of all Φ-filters (Definition 3.5.2). The next result is the dual of Proposition 3.5.3 and Corollary 3.5.5.

5.1.1 Proposition The function class $\Gamma(X,c\Phi)$ has the following properties:

i. $\Gamma(X,c\Phi) \supset \Phi$.

ii. $\Gamma(X,c\Phi)$ is composition closed.

iii. $c\Gamma(X,c\Phi) = c\Phi$.

iv. For two function classes Φ and Ψ we have $c\Phi = c\Psi$ if and only if $\Gamma(X,c\Phi) = \Gamma(X,c\Psi)$.

Proof: i. This follows by straightforward verification.

ii. This follows immediately from Proposition 3.2.4.

iii. The inclusion $c\Gamma(X,c\Phi) \subset c\Phi$ follows from (i) of this proposition. The other inclusion is exactly (i) of Proposition 3.5.3.

iv. This follows at once from (iii).

The next result is the dual of Proposition 3.5.10.

5.1.2 Proposition i. $\Gamma(X,c\Phi) = \Phi$ if and only if Φ is composition closed.

ii. $\Gamma(X,c\Phi)$ is the smallest composition closed function class on X containing Φ.

Proof: i. If $\Gamma(X,c\Phi) = \Phi$ then clearly Φ is composition closed, by Proposition 3.2.4. To prove the other implication, first note that by (i) of Proposition 5.1.1 we always have $\Phi \subset \Gamma(X,c\Phi)$. To prove the reverse inclusion we use the argument of A. Császár, in his proof of the equivalence of composition closedness and filterclosedness [31]. Suppose Φ is composition closed and let g be a function belonging to $\Gamma(X,c\Phi)$. Define $h : X \to \Pi_{\mu\in\Phi} \mathbb{R}_\mu$ by $x \to h(x) = (\mu(x))_{\mu\in\Phi}$, and let cl h(X) be the closure of h(X) in $\Pi_{\mu\in\Phi} \mathbb{R}_\mu$. For $z \in$ cl h(X) let $\underline{M}_z$ be the trace of the neighborhood filter $\underline{V}(z)$ to h(X). Then $h^{-1}(\underline{M}_z)$ is a filterbasis on X. Moreover, it generates a filter belonging to $c\Phi$. It follows that stack $g(h^{-1}(\underline{M}_z))$ is convergent. So we can define

$$k : \text{cl } h(X) \to \mathbb{R} : z \to k(z) = \lim \text{ stack } g(h^{-1}(\underline{M}_z)).$$

Straightforward calculations show that k is continuous and $g = k \circ h$. Since Φ is composition closed we finally can conclude that $g \in \Phi$.

ii. If Ψ is any composition closed function class, then applying (i) of this proposition we have

$$\phi \subset \Psi \Rightarrow c\phi \supset c\Psi$$

$$\Rightarrow \Gamma(X,c\phi) \subset \Gamma(X,c\Psi) = \Psi$$

In view of the preceding proposition we call $\Gamma(X,c\Phi)$ the <u>composition closed hull</u> of Φ.

From Propositions 3.2.4 and 5.1.2 (i) we now immediately have the following characterization of the classes $\Gamma(X,\underline{C})$:

5.1.3 Theorem A function class Φ on X is a collection $\Gamma(X,\underline{C})$ for some Cauchy structure $\underline{C}$ on X, if and only if Φ is composition closed.

5.1.4 Proposition If Φ is a collection $\Gamma(X,\underline{C})$ for some Cauchy structure $\underline{C}$ on X then $c\Phi$ is the coarsest Cauchy structure with this property.

Proof: This follow at once upon applying Proposition 3.5.3(i).

Let $\underline{A}$ be a point separating uniformly closed function algebra and lattice containing the constants and, as in Definition 4.3.12, let $\underline{H}(\underline{A})$ be its realcompact structure space. We prove the following result about the composition closed hull of $\underline{A}$.

5.1.5 Proposition The following properties hold:

i. $\Gamma(X,c\underline{A}) = C(\underline{H}(\underline{A}))/X$.
ii. $\underline{A}$ and $\Gamma(X,c\underline{A})$ have the same realcompact structure space.
iii. If $\underline{A}$ is inversion closed, then so is $\Gamma(X,c\underline{A})$.

Proof: i. Follows from the fact that $\underline{H}(\underline{A})$ is equivalent to the topology of the uniform completion of $(X,u\underline{A})$ with Cauchy subspace $(X,c\underline{A})$.

ii. Follows immediately from Proposition 5.1.1(iii).

iii. It is well known that inversion closedness of $\underline{A}$ implies that X is G_δ-dense in $\underline{H}(\underline{A})$. In view of (ii), then X is G_δ-dense in the strict completion $\underline{H}(\Gamma(X,c\underline{A}))$ of $(X,c\underline{A})$. Upon applying Propositions 5.1.1(iii) and 4.1.7, we can thus conclude that $\Gamma(X,c\underline{A})$ is inversion closed.

5.2 A NATURAL STRUCTURE ON Φ

Let Φ be a function class on X. We introduce a structure on Φ in the following way.

5.2.1 Definition $\gamma_\Phi = \{\Theta \in F(\Phi) \mid \text{stack } \Theta(\underline{F}) \text{ converges whenever } \underline{F} \text{ is a } \Phi\text{-filter}\}$

The following properties are shown by straightforward verification.

5.2.2 Proposition 1. (Φ,γ_Φ) is a Cauchy space.
2. γ_Φ is the coarsest Cauchy structure on Φ making the map $\Phi \times (X,c\Phi) \to R : (f,x) \to f(x)$ Cauchy continuous.

Let q_Φ be the convergence structure on Φ compatible with γ_Φ; then the following properties hold.

5.2.3 Proposition 1. Two Cauchy filters Θ and Θ' are equivalent in γ_Φ if and only if stack $\Theta(\underline{F})$ and stack $\Theta'(\underline{F})$ have the same limit whenever $\underline{F}$ is a Φ-filter on X.
2. Every Cauchy filter in γ_Φ has a pointwise limit.
3. A Cauchy filter Θ in γ_Φ converges for q_Φ if and only if its pointwise limit $\lim_p \Theta$ belongs to Φ.

Proof: (1) This follows at once from the fact that

$$\text{stack } (\Theta \cap \Theta')(\underline{F}) = \text{stack } \Theta(\underline{F}) \cap \text{stack } \Theta'(\underline{F})$$

(2) If $\Theta \in \gamma_\Phi$ then stack $\Theta(\dot{x})$ converges for every $x \in X$, since the principal ultrafilters are Φ-filters.

(3) This proof is analogous to the proof of Theorem 1.3 in [47]. For $\Theta \in \gamma_\Phi$ let $\lim_p \Theta$ be its pointwise limit. To prove the "if" part, suppose $\lim_p \Theta$ belongs to Φ and let $\underline{F}$ be any Φ-filter on X. One can easily verify that

$$cl_R\, \Theta\ (\underline{F}) < (\lim_p \Theta)\ (\underline{F})$$

It follows that stack $\Theta(\underline{F})$ and stack $(\lim_p \Theta)\ (\underline{F})$ have the same limit on R and consequently, stack $\Theta \cap (\lim_p \Theta)^{\cdot}\ (\underline{F})$ converges. From the arbitrariness of $\underline{F}$ we can conclude that $\Theta \cap (\lim_p \Theta)^{\cdot}$ belongs to γ_Φ.

To prove the "only if" part, suppose Θ converges in q_Φ to some function $h \in \Phi$. Then for every $x \in X$ we have

$$\text{stack } (\Theta \cap \dot{h})\ (\dot{x}) = \text{stack } \Theta(\dot{x}) \cap h(\dot{x})$$

It follows that stack $\Theta(\dot{x})$ converges to $h(x)$. So h is equal to the pointwise limit $\lim_p \Theta$.

The structure γ_Φ is compatible with the pointwise defined algebraic or lattice operations on Φ, as follows from the next result.

5.2.4 Proposition If Φ is a function algebra (a function lattice) and carries γ_Φ, then the algebraic (lattice) operations are all Cauchy continuous.

Proof: We give a proof for the addition:

$$s : \Phi \times \Phi \to \Phi$$

$$(f,g) \to s(f,g) = f + g$$

The other proofs are similar. Let $\Theta \in \gamma_\Phi$ and $\Theta' \in \gamma_\Phi$ and let $\underline{F}$ be a Φ-filter on X. Then the filter stack $\Theta(\underline{F})$ + stack $\Theta'(\underline{F})$ converges on R. Moreover we have:

$$s(\Theta \times \Theta)\ (\underline{F}) > \Theta(\underline{F}) + \Theta'(\underline{F})$$

Consequently stack $s(\Theta \times \Theta')$ belongs to γ_Φ.

5.2.5 Remark It is well known that the structure of uniform convergence is in general not compatible with the algebraic structure. For example, C(X) endowed with uniform convergence is neither a topological ring nor a topological vector space unless X is pseudocompact.

5.2.6 Corollary If Φ is a function algebra (function lattice) and $\Phi' \subset \Phi$ is a subalgebra (sublattice), then so is the closure of Φ' in (Φ, γ_Φ).

To further illustrate the meaning of γ_Φ we shall describe this structure in some special cases.

5.2.7 Example Let X be a realcompact topological space and let Φ = C(X). Then we have:

$\underline{F}$ is a Φ-filter $\iff$ F converges

Consequently, using the terminology of Section 3.4, we have,

$$\gamma_\Phi = \{\Theta \mid \text{stack } \Theta(\underline{F}) \text{ converges whenever } \underline{F} \text{ converges}\}$$
$$= \{\Theta \mid \Theta \text{ converges for continuous convergence}\}$$

5.2.8 Example Let $(X,\underline{C})$ be a domain (in particular let $\underline{C}$ be the compatible Cauchy structure of a uniformity on X, where X is a set of nonmeasurable cardinality). Further, we take $\Phi = \Gamma(X,\underline{C})$. Then we have

$$\underline{F} \text{ is a } \Phi\text{-filter} \iff F \in c\Gamma(X,\underline{C})$$
$$\iff \underline{F} \in \underline{C}$$

Consequently, with the notation of Section 1.3, we have

$$\gamma_\Phi = \{\Theta \mid \text{stack } \Theta(\underline{F}) \text{ converges whenever } \underline{F} \in \underline{C}\}$$
$$= \hat{\underline{C}}$$

In general, without the extra assumptions on X and $\underline{C}$ made in the previous example, the structures γ_Φ and $\hat{\underline{C}}$ on $\Phi = \Gamma(X,\underline{C})$ do not coincide, as follows from the next result.

5.2.9 Proposition If $(X,\underline{C})$ is a c^-embedded Cauchy space and $\Phi = \Gamma(X,\underline{C})$ then the following implication holds:

$$\gamma_\Phi = \hat{\underline{C}} \Rightarrow \underline{C} \text{ is a domain}$$

Proof: If $\underline{N}(X,\underline{C})$ stands for the natural completion of the c^-embedded space $(X,\underline{C})$ and $\Phi = \Gamma(X,\underline{C})$, we have the following implications:

$$\gamma_\Phi = \hat{\underline{C}} \Rightarrow \underline{N}(X,\underline{C}) \approx \underline{N}(X,c\Gamma(X,\underline{C}))$$
$$\Rightarrow \underline{C} = c\Gamma(X,\underline{C})$$

The relation between γ_Φ and the $\hat{}$-structure of Gazik and Kent is clarified somewhat further in the following statement, the proof of which follows by straightforward verification. For a function class Φ on X, as before let $c\Phi$ be equal to the collection of Φ-filters and $c\hat{}\Phi$ be equal to $\{\Theta \in F(\Gamma(X,c\Phi)) \mid \text{stack } \Theta(\underline{F})$ converges whenever $\underline{F} \in c\Phi\}$.

5.2.10 Proposition For any function class Φ on X, the space (Φ, γ_Φ) is a Cauchy subspace of $(\Gamma(X,c\Phi),c\hat{}\Phi)$.

From Examples 5.2.7 and 5.2.8 it follows that the Cauchy structure γ_Φ is in general not uniformizable, and not even a topological or pretopological Cauchy structure on Φ. However, from Proposition 5.2.10 and the results on the $\hat{}$-structure developed by Gazik and Kent in [47] it follows that γ_Φ- has the following nice properties.

5.2.11 Proposition If Φ is any function class on X then (Φ,γ_Φ) is pseudotopological μ-regular and Cauchy separated.

Proof: It follows from the results of [47] that $(\Gamma(X,c\Phi),\hat{c}\Phi)$ is a c-embedded convergence space. By Theorem 2.4.4 we can conclude that (Φ, γ_Φ), being a Cauchy subspace of $(\Gamma(X,c\Phi),\hat{c}\Phi)$, is itself $c\hat{}$-embedded. Hence it is μ-regular, Cauchy separated, and pseudotopological.

When the function class Φ is embedded in $C(X,t\Phi)$ it inherits the Cauchy structure $\underline{C}_u$, compatible with the uniformity of uniform convergence, and also it inherits from $C(X,t\Phi)$ the Cauchy structure $\underline{C}_{\underline{G}_c}$ of the uniform convergence structure $\underline{G}_c$ of continuou convergence, as defined in Section 3.4. We shall compare the structures induced on Φ with the Cauchy structure γ_Φ.

5.2.12 Proposition The structure γ_Φ is coarser than the Cauchy structure $\underline{C}_u$ of uniform convergence.

Proof: Let $\Theta \in \underline{C}_u$ and let $\underline{F}$ be a Φ-filter on X. We show that stack $\Theta(\underline{F})$ is a Cauchy filter on the real line. For an arbitrary entourage V on R we choose an entourage U such that $U^3 \subset V$. Let $\underline{B} \in \Theta$ be such that for every f,g in $\underline{B}$ we have

$$(f(x),g(x)) \in U \qquad \text{for every } x \in X$$

Fix $h \in B$ and choose $F \in \underline{F}$ satisfying

$$h(F) \times h(F) \subset U$$

It follows that $\underline{B}(F) \in \Theta(\underline{F})$ and that, moreover,

$$\underline{B}(F) \times \underline{B}(F) \subset V$$

5.2.13 Remark In general γ_Φ can be strictly coarser than the Cauchy structure of uniform convergence. For example, take X = R with the usual complete Cauchy structure and let Φ be the collection of (Cauchy) continuous maps. The Fréchet filter of the sequence

$$f_n(x) = \frac{1}{n}x \qquad n \geq 1,\ x \in R$$

belongs to γ_Φ but does not belong to $\underline{C}_u$.

On function classes of bounded functions, however, the structures $\underline{C}_u$ and γ_Φ coincide. This will be discussed in Section 5.4.

5.2.14 Proposition The structure γ_Φ is finer than the Cauchy structure $\underline{C}_{\underline{G}_c}$ of continuous convergence.

Proof: Let $\Theta \in \gamma_\Phi$ and let $\underline{F}$ be a filter on X which $t\Phi$-converges. Clearly then $\underline{F}$ is a Φ-filter on X and consequently stack $\Theta(\underline{F})$ converges. It follows that Θ belongs to the Cauchy structure on Φ induced by the continuous convergence on $C(X, t\Gamma(X))$.

5.2.15 Remark In general γ_Φ can be strictly finer than the structure induced by $\underline{C}_{\underline{G}_c}$. For example, take $X = (0,1]$ with the usual Cauchy structure $\underline{C}$ induced by its extension $[0,1]$ and let $\Phi = \Gamma(X, \underline{C})$. The Fréchet filter Θ of the sequence

$$f_n(x) = \begin{cases} 0 & \text{for } x \geq \frac{1}{n} \\ n - n^2 x & \text{for } x < \frac{1}{n} \end{cases} \qquad n \geq 1$$

is continuously convergent on $C(X, t\Phi)$. Consequently Θ belongs to $\underline{C}_{\underline{G}_c}$. On the other hand however, the trace $\underline{M}$ of the usual neighborhood of 0 on $[0,1]$ is a Φ-filter and stack $\Theta(\underline{M})$ does not converge. Consequently Θ does not belong to γ_Φ.

5.3 COMPLETE FUNCTION CLASSES

5.3.1 Definition A function class $\Phi \in R^X$ is <u>complete</u> if (Φ, γ_Φ) is a complete Cauchy space.

The following characterization follows immediately from Proposition 5.2.3.

5.3.2 Proposition A function class Φ is complete if and only if $\lim_p \Theta$ belongs to Φ whenever Θ belongs to γ_Φ.

5.3.3 Proposition For any function class Φ on X completeness for γ_Φ is stronger than uniform completeness and weaker than completeness for continuous convergence.

Proof: First suppose (Φ, γ_Φ) is complete and let Θ be a filter on Φ which is a uniform Cauchy filter. By Proposition 5.2.12 the filter Θ belongs to γ_Φ. Then applying Proposition 5.3.2, the function $f = \lim_p \Theta$ belongs to Φ. It follows that Θ is uniformly convergent to f. Finally, we can conclude that $(\Phi, \underline{C}_u)$ is complete. To prove the second assertion, now suppose Φ is complete for $\underline{C}_{G_c}$ and let $\Theta \in \gamma_\Phi$. By Proposition 5.2.14, Θ is then a Cauchy filter for $(\Phi, \underline{C}_{G_c})$. It follows that Φ is continuously convergent to a function $f \in \Phi$. But then $\lim_p \Theta = f \in \Phi$ and by Proposition 5.3.2 finally (Φ, γ_Φ) is complete.

5.3.4 Proposition Every collection $\Gamma(X)$ is complete.

Proof: This follows either from the application of Propositions 5.2.10 and 3.5.3 combined with the result of Gazik and Kent which we recalled in Proposition 1.4.15, or it can be shown directly using the following implications: Let $\Phi = \Gamma(X)$. Then

$$\begin{aligned} \Theta \in \gamma_\Phi &\Rightarrow (\lim_p \Theta)(\underline{F}) > cl_R \Theta(F), \text{ whenever } \underline{F} \text{ is a } \Phi\text{-filter} \\ &\Rightarrow \text{stack } (\lim_p \Theta)(\underline{F}) \text{ converges, whenever } \underline{F} \text{ is a } \Phi\text{-filter} \\ &\Rightarrow \lim_p \Theta \in \Gamma(X, c\Phi) \end{aligned}$$

Since Φ is composition closed, Proposition 5.1.2 (i) now implies that $\lim_p \Theta \in \Phi$. In view of Proposition 5.3.2 this implies that Φ is complete.

As we already pointed out in Section 3.4 and in Proposition 4.1.5 a function class $\Gamma(X)$ is generally not $\underline{C}_{G_c}$-complete (unless it is a class of type C(X)). So as a corollary of the previous proposition, we can conclude that the new completeness notion is generally strictly weaker than completeness for continuous convergence.

We now give an example to show that the new completeness notion is generally strictly stronger than uniform completeness. For this purpose we use an example studied in detail by A. Császár in [31].

5.3.5 Example Let R be the real line and for Φ we take the collection of all functions of first Baire class. It is well known that this class is a point separating uniformly closed function algebra containing the constants. From Theorem 2.5. in [31], moreover, we have

$$c\Phi = \{\dot{x} \mid x \in \mathrm{R}\}$$

There exists a sequence $\{f_n \mid n \in \mathrm{N}\}$ of functions in Φ, with a pointwise limit f that does not belong to Φ. Since stack $\langle f_n \rangle (\dot{x})$ converges for all $x \in \mathrm{R}$, we can conclude that $\langle f_n \rangle \in \gamma_\Phi$. So by Proposition 5.3.2, Φ is not γ_Φ-complete.

Next we come to our main theorem, the characterization of the function classes of type $\Gamma(X)$ by means of the new completeness notion. First we need the following lemma.

5.3.6 Lemma For a function class Φ on X we introduce the following notation:

$u\Phi$ is the initial uniformity on X determined by Φ.
$(\hat{X}, (u\Phi)\hat{})$ is the uniform completion of $(X, u\Phi)$.
$(\hat{X}, \hat{\tau})$ is the compatible topological space of $(\hat{X}, (u\Phi)\hat{})$.
$\hat{\Phi} = \{\hat{f} \mid f \in \Phi\}$ where $\hat{f}$ is the continuous extension of f to $(\hat{X}, \hat{\tau})$.
$u\hat{\Phi}$ is the initial uniformity on $\hat{X}$ determined by $\hat{\Phi}$.

With these notations we now have

$$(u\Phi)\hat{} = u\hat{\Phi}$$

Proof: The functions in $\hat{\Phi}$ are all uniformly continuous on $(\hat{X}, (u\Phi)\hat{})$. Consequently we clearly have $u\hat{\Phi} \leq (u\Phi)\hat{}$. To prove the other inequality, let U be a basic entourage in $u\Phi$ of the form

$$U = \{(x,y) \mid |f_i(x) - f_i(y)| < \varepsilon,\ i = 1,\ldots,n\}$$

where $f_1,\ldots,f_n$ belong to Φ and $\varepsilon > 0$. Let $\hat{U}$ be the corresponding entourage in $(u\Phi)\hat{}$. If $\underline{M}$ and $\underline{N}$ denote minimal Cauchy filters for the uniformity $u\Phi$ then the set

$$W = \{(\underline{M},\underline{N}) \mid |\hat{f}_i(\underline{M}) - \hat{f}_i(\underline{N})| < \frac{\varepsilon}{2},\ i = 1,\ldots,n\}$$

is a basic entourage for $u\hat{\Phi}$.

To show that W is contained in $\hat{U}$, let $\underline{M}$ and $\underline{N}$ be $u\Phi$-minimal Cauchy filters, and for $i \in \{1,\ldots,n\}$, let a_i be the limit of stack $f_i(\underline{M})$. Then we have

$$|\hat{f}_i(\underline{M}) - \hat{f}_i(\underline{N})| < \frac{\varepsilon}{2} \Rightarrow \left(a_i - \frac{\varepsilon}{2},\ a_i + \frac{\varepsilon}{2}\right)$$
$$\in \text{stack } f_i(\underline{M}) \cap \text{stack } f_i(\underline{N})$$

Consequently, for $i \in \{1,\ldots,n\}$ we can choose sets

$$A_i \in \underline{M} \quad \text{with } f_i(A_i) \subset \left(a_i - \frac{\varepsilon}{2},\ a_i + \frac{\varepsilon}{2}\right)$$

and

$$B_i \in \underline{N} \quad \text{with } f_i(B_i) \subset \left(a_1 - \frac{\varepsilon}{2},\ a_i + \frac{\varepsilon}{2}\right)$$

Finally, if we put

$$A = \bigcap_{i=1}^{n} A_i \qquad B = \bigcap_{i=1}^{n} B_i$$

then clearly $A \cup B$ is U-small and belongs to $\underline{M} \cup \underline{N}$.

5.3.7 Theorem A function class on X is of type $\Gamma(X)$, for a suitable Cauchy structure on X, if and only if it is a complete function algebra containing the constants.

Proof: The "if" part was shown in Propositions 5.3.4 and 3.2.1. To prove the "only if" part, let (Φ, γ_Φ) be a complete function algebra containing the constants. We use the same notations as in the previous lemma, to which we add, as in Section 5.2,

$$c\hat{\ }\Phi = \{\Theta \in F(\Gamma(X,c\Phi)) \mid \text{stack } \Theta(\underline{F}) \text{ converges whenever } \underline{F} \in c\Phi\}$$

It is clear that $\hat{\Phi} \subset C(\hat{X}, \hat{\tau})$. We show that this inclusion in fact is an equality. Since the continuous extensions of the maps in Φ are unique, it follows that $\hat{\Phi}$ is a subalgebra of $C(\hat{X}, \hat{\tau})$ containing the constants. Let κ be the map defined in Section 3.2:

$$\kappa : \Gamma(X,c\Phi) \to C(\hat{X}, \hat{\tau})$$
$$f \to \kappa(f) = \hat{f}$$

When $\Gamma(X,c\Phi)$ is endowed with the Cauchy structure $c\hat{\ }\Phi$ and $C(\hat{X}, \hat{\tau})$ is endowed with $\underline{C}_{\underline{G}_c}$, then the result of Gazik and Kent, which we formulated in Proposition 3.4.8, states that κ is a Cauchy isomorphism. In view of Proposition 5.2.10 the space (Φ, γ_Φ) is a complete Cauchy subspace of $(\Gamma(X,c\Phi), c\hat{\ }\Phi)$. So it follows that $\kappa(\Phi) = \hat{\Phi}$ is closed in $C(\hat{X}, \hat{\tau})$ for continuous convergence.

By Lemma 5.3.6 we have

$$(u\Phi)^{\wedge} = u^{\wedge}\Phi$$

which then implies

$$\hat{\tau} = t\hat{\Phi}$$

Moreover, by Theorem 48 in [21], $\hat{\Phi}$, being an algebra closed for continuous convergence, is also a uniformly closed lattice and algebra. It is well known (see, for instance, Isbell's paper [61]) that then $\hat{\Phi}^*$, the collection of bounded functions in $\hat{\Phi}$, determines the same initial topology as $\hat{\Phi}$. So we have

$$\hat{\tau} = t\hat{\Phi}^*$$

Finally, all conditions on $\hat{\Phi}$ for the application of the result of Binz, formulated in Lemma (3.4.9), are fulfilled. So we can conclude that

$$\hat{\Phi} = C(\hat{X}, \hat{\tau})$$

and taking the restrictions to X we obtain the equality

$$\Phi = \Gamma(X, c\Phi)$$

In [32] A. Császár posed the following question: Is it possible to characterize the composition closed function classes by some finite composition property (for instance, being an algebra containing the constants) and a suitable approximation property? Combining the results of Theorems (5.1.3) and 5.3.7 we now can answer this question affirmatively.

5.3.8 Proposition A function class Φ is composition closed if and only if it is a complete function algebra containing the constants.

5.4 BOUNDED FUNCTIONS

In this section we investigate the particular situation of function classes consisting of bounded functions. In view of Lemma 3.3.1 we immediately have the following result.

5.4.1 Proposition For a function class Φ on X the following properties are equivalent:

1. Φ consists of bounded functions.
2. $c\Phi$ is totally bounded.
3. $\Gamma(X, c\Phi)$ consists of bounded functions.

Although in general γ_Φ can be strictly coarser than $\underline{C}_u$, as was pointed out in Remark 5.2.13, it will follow from the next result that this is not so for the bounded case.

5.4.2 Proposition If Φ is a function class of bounded functions, then γ_Φ coincides with the Cauchy structure induced on Φ by the Cauchy structure $\underline{C}_u$ of uniform convergence.

Proof: Suppose Φ consists of bounded functions. Let $\hat{\tau}$ be the topology compatible with the uniform completion $(\hat{X}, (u\Phi)\hat{})$. Since $c\Phi$ is totally bounded, again using the notation of the proof of Theorem 5.3.7, we have that $(\hat{X}, \hat{\tau})$ is compact. It is well known that this implies that the structures $\underline{C}_{\underline{G}_c}$ and $\underline{C}_u$ on $C(\hat{X}, \hat{\tau})$ coincide. Upon applying Propositions 3.4.1 and 3.4.8 we now have that in the following diagram the maps κ, id and κ^{-1} are all Cauchy isomorphisms.

$$\begin{array}{ccc} \Gamma(X, c\Phi),\ c\hat{}\Phi & \xrightarrow{\kappa} & C(\hat{X}, \hat{\tau}),\ \underline{C}_{\underline{G}_c} \\ & & \downarrow \text{id} \\ \Gamma(X, c\Phi), \underline{C}_u & \xleftarrow{\kappa^{-1}} & C(\hat{X}, \hat{\tau}),\ \underline{C}_u \end{array}$$

It follows that the structures $c^\wedge\Phi$ and $\underline{C}_u$ on $\Gamma(X,c\Phi)$ are isomorphic. Taking the induced structures on Φ we are done.

When we apply our main theorem (5.3.7) to a bounded function class then we obtain the following result.

5.4.3 Theorem If Φ is a function class of bounded functions on X, then the following properties are equivalent:

1. Φ is a uniform closed function algebra containing the constants.
2. Φ is a collection $\Gamma(X)$ for a suitable Cauchy structure on X.
3. Φ is a collection $U(X,\underline{U})$ of real-valued uniformly continuous maps for a suitable uniformity $\underline{U}$ on X.
4. Φ is composition closed.

Proof: (1) $\Rightarrow$ (2) If Φ is a uniformly closed function algebra containing the constants, then by Proposition 5.4.2, Φ is also γ_Φ-complete. Theorem 5.3.7 then implies (2).

(2) $\Rightarrow$ (3) Suppose $\Phi = \Gamma(X)$. For the uniformity $u\Phi$ we now have $\Phi \subset U(X,u\Phi) \subset \Gamma(X,c\Phi) \subset \Phi$.

(3) $\Rightarrow$ (4) Suppose $\Phi = U(X,\underline{U})$ for some uniformity $\underline{U}$ on X. We take again the initial uniformity $u\Phi$ on X. Since $u\Phi$ is totally bounded we have

$$\begin{aligned}\Gamma(X,c\Phi) &= U(X,u\Phi)\\ &\subset U(X,\underline{U})\\ &= \Phi.\end{aligned}$$

By Proposition 5.1.2 it follows that Φ is composition closed.

(4) $\Rightarrow$ (1) By Proposition 5.3.8, composition closedness of Φ implies that Φ is a γ_Φ-complete function algebra containing the constants. But then it is also uniformly complete.

5.4.4 Remark The equivalence of (1) and (2) also follows directly from the results of Isbell [61] and Hendriksen and Johnson [54] formulated in Proposition 4.3.2, stating that a collection Φ of bounded functions satisfying the assumptions of (1) is equal to $C(\underline{H}(\Phi))/X$. It follows that for the Cauchy structure induced by $\underline{H}(\Phi)$ on X we have $\Phi = \Gamma(X)$.

5.4.5 Remark For a uniformly closed function algebra Φ containing the constants and consisting of bounded functions, whether $\Phi = \Gamma(X,c\Phi)$ or $\Phi = C^*(X,t\Phi)$ depends on the separation properties of Φ. Indeed,

$\Phi = \Gamma(X,c\Phi)$ holds if Φ is point separating (Theorem 5.4.3).
$\Phi = C^*(X,t\Phi)$ holds if Φ separates the zero sets of $t\Phi$ and is point separating.

5.5 THE FUNCTION CLASS Φ^*

In this section, Φ is an arbitrary function class on X and Φ^* is the collection of its bounded functions. To be able to apply the theory of the previous sections to Φ^*, we have to make the assumption that Φ^* is point separating. In general this property does not follow from the point separability of Φ. For instance, if Φ is the collection of real polynomials on R then Φ^* contains nothing but the constants. However, it is well known from Isbell's paper [61] that if Φ is a function algebra containing the constants, which is closed under bounded inversion, then the point separability of Φ implies the point separability of Φ^*. So, in particular, for a uniformly closed function algebra and lattice containing the constants, and a fortiori for a composition closed function class, we do not need to impose an extra point separability condition on Φ^*. For general function classes, however, the extra assumption is made without further explicit mention.

From Corollary 3.3.6 it is immediately clear that Φ^* is composition closed whenever Φ is composition closed and so we clearly have the following result.

5.5.1 Proposition If Φ is composition closed then we have

$$\Gamma^*(X,c\Phi) = \Gamma(X,c\Phi^*)$$

5.5.2 Remark For an arbitrary function class Φ on X the inclusion

$$\Gamma^*(X,c\Phi) \supset \Gamma^*(X,c\Phi^*)$$

always holds. It can, however, be a strict inclusion. An example is provided by taking for Φ the collection of functions of first Baire class on the real line R, as in Example 5.3.5. As we saw, this class Φ is a point separated uniformly closed algebra containing the constants and

$$c\Phi = \{\dot{x} \mid x \in \mathsf{R}\}$$

So clearly

$$\Gamma^*(X,c\Phi) = (\mathsf{R}^{\mathsf{R}})^*$$

The collection Φ^* is also a uniformly closed point separating algebra containing the constants. In view of Theorem 5.4.3 we can conclude that Φ^* is composition closed. Consequently we have

$$\Gamma(X,c\Phi^*) = \Phi^*$$

Necessary and sufficient conditions for the equality of $\Gamma(X,c\Phi^*)$ and $\Gamma^*(X,c\Phi)$ are given in the next result.

5.5.3 Proposition Let Φ be a function class on X, let $(\hat{X},\hat{\tau})$ be the topological space of the uniform completion of $(X,u\Phi)$ and let $(\hat{X}^*,\hat{\tau}^*)$ be the topological space of the uniform completion of $(X,u\Phi^*)$. Then the following are equivalent:

1. $\beta(\hat{X},\hat{\tau}) = (\hat{X}^*,\hat{\tau}^*)$
2. $\Gamma^*(X,c\Phi) = \Gamma(X,c\Phi^*)$

Proof: (1) $\Rightarrow$ (2) The only nontrivial inclusion is $\Gamma^*(X,c\Phi) \subset \Gamma(X,c\Phi^*)$. To prove this, let $f \in \Gamma^*(X,c\Phi)$. There exists a bounded function g in $C(\hat{X},\hat{\tau})$ that extends f. From assumption (1) it follows that there exists a bounded function h in $C(\hat{X}^*,\hat{\tau}^*)$ extending g. Clearly then, the restriction of h to X equals f and belongs to $\Gamma(X,c\Phi^*)$.
(2) $\Rightarrow$ (1) Clearly $\beta(\hat{X},\hat{\tau})$ is a compactification of $(X,t\Phi)$ satisfying

$$C(\beta(\hat{X},\hat{\tau}))/X = C^*(\hat{X},\hat{\tau})/X$$

$$= \Gamma^*(X,c\Phi)$$

So under the assumption (2) we have

$$C(\beta(\hat{X},\hat{\tau}))/X = \Gamma(X,c\Phi^*)$$

Upon applying Proposition 4.3.2 to the algebra $\underline{A} = \Gamma(X,c\Phi^*)$ it follows that

$$\beta(\hat{X},\hat{\tau}) \approx \underline{H}(\underline{A}^*) \approx (\hat{X}^*,\hat{\tau}^*).$$

Next we consider Φ^* as a subset of Φ. As we know from Propositions 5.2.12 and 5.2.14 we have

$$q_c \leq q_\Phi \leq q_u$$

if q_c, q_Φ, q_u are the convergence structures on Φ compatible with the Cauchy structures $\underline{C}_{\underline{G}_c}$, γ_Φ, and $\underline{C}_u$, respectively. It is well known that for (Φ, q_u), the subset Φ^* is closed, and that if Φ is a lattice containing the constants, then the subset Φ^* is dense in (Φ, q_c). For the structure q_Φ, lying in between the two others, we have the following result:

5.5.4 Proposition If Φ is a function lattice containing the constants then Φ^* is a dense subset of (Φ, q_Φ).

Proof: For $f \in \Phi$ and $n \in \mathbb{N}$ let $f_n = f \vee (-n) \wedge n$ and let Θ be the Fréchet filter of this sequence. For any Φ-filter $\underline{F}$ on X the filter stack $f(\underline{F})$ converges on $\mathbb{R}$. Moreover, it is clear that stack $\Theta(\underline{F})$ then also converges to the same limit. Consequently we have $\Theta \cap \dot{f} \in \gamma_\Phi$.

6

Compatible Cauchy Spaces with a Given Collection of Cauchy Continuous Maps

In this chapter the theory of Cauchy spaces and Cauchy continuous maps will be applied to find solutions for the following general problem (T) posed by A. Császár in [33].

> Let X be a topological space and Φ a class of real-valued functions on X. Given a class T of topological spaces, find necessary and/or sufficient conditions that there exists an extension Y of X, that belongs to T, and such that Φ consists of those functions that have a continuous extension to Y.

In [32] Császár gives a necessary and sufficient condition for the case when T is the class of all completely regular spaces. Later, in [33], he establishes necessary and sufficient conditions for the class T of all topological spaces (without any separation assumption) and for the class of all T_1 topological spaces.

To treat the problem for the class of all Hausdorff topological spaces, Császár introduced the following conditions for a function class Φ on a topological space X:

If $f \in C(X)$ and stack $f(\underline{F})$ converges whenever $\underline{F}$ is an open Φ-filter with empty adherence, then $f \in \Phi$. (B)

If $f \in C(X)$ and stack $f(\underline{F})$ converges whenever $\underline{F}$ is a $u\Phi$-round Φ-filter with empty adherence, then $f \in \Phi$. (C)

He has shown that for the problem (T) in the class T of all Hausdorff topological spaces, (C) is a sufficient condition that is not necessary, whereas (B) is a necessary condition. Whether (B) is sufficient was formulated as a question.

In [87] we treated a problem (Q) analogous to problem (T) in the framework of convergence spaces (without the assumption of any separation axioms) and we gave solutions for problem (Q) in the class Q of all reciprocal convergence spaces and in the class of all relatively ω-regular reciprocal convergence spaces. As corollaries to these results we derived a solution for Császár's problem (T), for the class T of all relatively ω-regular reciprocal topological spaces.

In this chapter we shall develop solutions of problem (Q) in the framework of Hausdorff convergence spaces. As in [87] our solutions are obtained by means of an application of the theory of Cauchy spaces and Cauchy continuous maps. First we give a formulation of problems (T) and (Q) in terms of Cauchy spaces.

By this method we then give a solution to problem (Q) in the class Q of all Hausdorff convergence spaces. For this purpose the following condition is introduced.

If $f \in C(X)$ and stack $f(\underline{F})$ converges whenever $\underline{F}$ is a Φ-filter with empty adherence, then $f \in \Phi$. (D)

We prove that condition (D) is necessary and sufficient. As a corollary it follows that (D) is a necessary condition for problem (T) in the class of all Hausdorff topological spaces. We show that (D) is not sufficient for this class.

We make a comparison between conditions (B), (C), and (D). It appears that (C) implies (B) and that (B) implies (D). In general, none of these implications is reversible. On a regular Hausdorff topological space, however, (B) and (D) are equivalent, and on a locally compact Hausdorff topological space the properties (B), (C), and (D) are all equivalent.

Further, we give a solution to problem (Q) in the class Q of all relatively ω-regular Hausdorff convergence spaces. Again, as a corollary we obtain solutions for the corresponding class T. The meaning of Császár's condition (C) defined above is illustrated in this context.

Throughout this chapter, consistently with the previous chapters, "convergence space" or "topological space" stand for Hausdorff convergence space or Hausdorff topological space, respectively.

6.1 THE CAUCHY FORMULATIONS OF PROBLEMS (Q) and (T)

To give a formulation of problem (Q) we need the following concept.

6.1.1 Definition An extension Y of a convergence space is an R-<u>extension</u> if every Cauchy continuous real-valued map on X with the Cauchy structure induced by Y has a continuous extension to Y.

As we already know from Theorem 1.4.12, every strict extension is an R-extension and then a fortiori, every topological extension is an R-extension. However, an R-extension does not need to be strict. For instance, a c-embedded extension always is an R-extension, as follows from Proposition 1.4.16, but it is not necessarily strict.

Let X be a set with a convergence structure on it, and let Φ be a function class on the set X. Let Q be a class of convergence spaces. Then problem (Q) is formulated as follows:

> Find necessary and/or sufficient conditions that there exists an R-extension Y of X that belongs to Q and such that Φ consists of those functions that have a continuous extension to Y.

To give a formulation of this problem in terms of Cauchy structures we introduce the following concept.

6.1.2 Definition A Cauchy space $(X,\underline{C})$ is a Q-Cauchy space if it has a completion that belongs to Q and is an R-extension.

The proof of the following equivalence, which gives a formulation of problem (Q) by means of Cauchy structures, is a straightforward verification.

6.1.3 Proposition The following properties are equivalent:

1. There exists an R-extension Y of X belonging to Q and such that $C(Y)/X = \Phi$.
2. There exists a compatible Q-Cauchy structure $\underline{C}$ on X such that $\Gamma(X,\underline{C}) = \Phi$.

If T is a class of topological spaces we can always consider a corresponding class Q of convergence spaces such that T = $Q \cap \tau$, where τ is the class of all topological convergence spaces. Clearly then the formulation of problem (T) is exactly the same as the formulation of the generalized problem $Q \cap \tau$. In view of this fact we now have the next result.

6.1.4 Proposition The following properties are equivalent

1. There exists a (topological) extension Y of X belonging to T and such that $C(Y)/X = \Phi$.
2. There exists a compatible T Cauchy structure $\underline{C}$ on X such that $\Gamma(X,\underline{C}) = \Phi$.

With respect to condition (2) in the previous proposition the following remark is in order.

6.1.5 Remark If τ is the class of all topological spaces, a Cauchy structure that is compatible with a topological structure on X does not need be a τ-Cauchy structure. Moreover, if Q is a class of convergence spaces then a Cauchy space $(X,\underline{C})$ can be a Q-Cauchy space and a τ-Cauchy space without being a $Q \cap \tau$-Cauchy space. Many examples of this situation occur in the literature. See, for example, [37] and [86].

6.2 EXTENSIONS IN THE CLASS OF ALL CONVERGENCE SPACES

In this section, (X,q) is a convergence space and Φ is a function class on X, and we suppose they are related by the assumption

$$\Phi \subset C(X,q)$$

Moreover, Q is the class of all convergence spaces.

We adopt the usual notation $c\Phi$ for the collection of all Φ-filters. Note, however, in this context, since we do not assume

Φ to be point separating, that $c\Phi$ is not a T_1 Cauchy space. The following obvious result will be used several times.

6.2.1 Lemma If $\underline{D}$ is a Cauchy structure on a convergence space (X,q), containing the q-convergent filters, and $\underline{C} \leq \underline{D}$ for some compatible Cauchy structure $\underline{C}$, then $\underline{D}$ itself is compatible.

6.2.2 Definition

$d\Phi = \{\underline{F} \in F(X) \mid \underline{F}$ is q-convergent or $\underline{F} \in c\Phi,\ \alpha_q\underline{F} = \emptyset\}$

The next result follows immediately. In the formulation, $\underline{C}_q$ stands for the coarsest Cauchy structure compatible with q [70]:

$$\underline{C}_q = \{\underline{F} \in F(X) \mid \underline{F} \text{ q-converges or } \alpha_q\underline{F} = \emptyset\}$$

6.2.3 Propositions The following properties hold:

1. $d\Phi$ is a T_1 Cauchy structure on X.
2. $d\Phi = c\Phi \cap \underline{C}_q$.
3. $d\Phi$ is compatible with (X,q).

6.2.4 Proposition If there exists a compatible Cauchy structure $\underline{C}$ on (X,q) such that $\Gamma(X,\underline{C}) = \Phi$, then $d\Phi$ is the coarsest structure with these properties.

Proof: Let $\underline{C}$ be a Cauchy structure compatible with (X,q) and satisfying $\Gamma(X,\underline{C}) = \Phi$. The compatibility assumption implies that $\underline{C} \subset \underline{C}_q$ and the second assumption implies $\underline{C} \subset c\Phi$ in view of the results of Section 5.1. So by application of Proposition 6.2.3 we clearly have $d\Phi \leq \underline{C}$. Moreover upon applying Theorem 5.1.3 we have

$$\Gamma(X,\underline{C}) \supset \Gamma(X,d\Phi)$$
$$\supset \Gamma(X,c\Phi)$$
$$= \Phi$$

and we are done.

Next we recall the definition given above of condition (D) for a function class Φ.

6.2.5 Definition If $f \in C(X,q)$ and stack $f(\underline{F})$ converges whenever $\underline{F}$ is a Φ-filter with empty q-adherence, then $f \in \Phi$. (D)

Next we come to our main theorem, which in view of Proposition 6.1.3 is a solution of problem (Q).

6.2.6 Theorem The following properties are equivalent:

1. There exists a compatible Q-Cauchy structure $\underline{C}$ on X such that $\Gamma(X,\underline{C}) = \Phi$.
2. There exists a compatible Cauchy structure $\underline{C}$ on (X,q) such that $\Gamma(X,\underline{C}) = \Phi$.
3. Φ satisfies (D).
4. Φ is composition closed and determines the same initial topology as $\Gamma(X,d\Phi)$.

Proof: (1) $\Rightarrow$ (2) This is trivial.

(2) $\Rightarrow$ (3) Suppose $\underline{C}$ is compatible and $\Gamma(X,\underline{C}) = \Phi$. Then in view of Proposition 6.2.4 we have the following implications:

$$f \in C(X,q) \text{ and stack } f(\underline{F}) \text{ converges } \forall\ \underline{F} \in c\Phi,\ \alpha_q\underline{F} = \emptyset$$
$$\Rightarrow f \in \Gamma(X,d\Phi)$$
$$\Rightarrow f \in \Phi$$

(3) $\Rightarrow$ (4) Suppose Φ satisfies condition (D). Applying Propositions 5.1.1 (i) and 6.2.3 (2) we have

$$\begin{aligned}\Phi &\subset \Gamma(X,c\Phi)\\ &\subset \Gamma(X,d\Phi)\\ &\subset \Phi\end{aligned}$$

In view of Proposition 5.1.2, then Φ is composition closed. Moreover, Φ determines the same initial topology as $\Gamma(X,d\Phi)$.

(4) $\Rightarrow$ (1) Suppose Φ is composition closed and determines the same initial topology as $\Gamma(X,d\Phi)$. Clearly, in view of Proposition 6.2.3, we have

$$\Gamma(X,c\Phi) \subset \Gamma(X,d\Phi)$$

To prove the other inclusion, let $f \in \Gamma(X,d\Phi)$ and let $\underline{F} \in c\Phi$. Either $\underline{F}$ $t\Phi$-converges or its $t\Phi$-adherence is empty. In the first case we can conclude that stack $f(\underline{F})$ converges, using the $t\Phi$-continuity of f, which follows from the equality $t\Phi = t\Gamma(X,d\Phi)$. In the second case the q-adherence of $\underline{F}$ also has to be empty and consequently $\underline{F} \in d\Phi$, and so we again can conclude that stack $f(\underline{F})$ converges. Finally, we have shown that $f \in \Gamma(X,c\Phi)$ and from the arbitrariness of f we can conclude that

$$\Gamma(X,c\Phi) = \Gamma(X,d\Phi)$$

Applying Proposition 5.1.2 we can conclude that

$$\Gamma(X,d\Phi) = \Phi$$

Moreover, by Proposition 6.2.3 (3), $d\Phi$ is compatible. Since every Cauchy structure is a Q-Cauchy structure, we are done.

We now investigate how the previous theorem is related to our earlier work in [87]. We show that the Cauchy structure $d\Phi$ coincides with the collection $\underline{D}_\Phi$ introduced in [87].

6.2.7 Proposition

$d\Phi = \{\underline{F} \in F(X) \mid \underline{F} \in c\Phi$, and $\forall\ x \in X$, if $\underline{F}$ $t\Phi$-converges to x, then either $\underline{F}$ q-converges to x or $x \notin \alpha_q\underline{F}\}$

Proof: Suppose $\underline{F} \in d\Phi$. Then we already have $\underline{F} \in c\Phi$. Now two cases can occur. Either $\alpha_q\underline{F} = \emptyset$ and then clearly $\underline{F}$ belongs to the collection on the right, or $\underline{F}$ q-converges to some $x \in X$. In this case, suppose $\underline{F}$ $t\Phi$-converges to some point y. If $y = x$ then $\underline{F}$ q-converges to y. If $y \neq x$ then clearly $y \notin \alpha_q\underline{F}$. So we again can conclude that $\underline{F}$ belongs to the collection on the right.

To prove the other inclusion, suppose F belongs to the collection on the right. Then we already have $\underline{F} \in c\Phi$. To prove that $\underline{F} \in \underline{C}_q$, suppose $x \in \alpha_q\underline{F}$. Then we clearly have $x \in \alpha_{t\Phi}\underline{F}$, and since $\underline{F}$ is a Φ-filter this implies that $\underline{F}$ $t\Phi$-converges to x. Consequently, $\underline{F}$ q-converges to x and we are done.

6.3 EXTENSIONS IN THE CLASS OF ALL PSEUDOTOPOLOGICAL SPACES

In this section (X,q) again is a convergence space and Φ is a function class on X related to (X,q) by

$$\Phi \subset C(X,q)$$

Moreover in this section, **Q** is the class of all pseudotopological convergence spaces. Further, let $d\Phi$ and condition (D) be defined as in the previous section. To solve problem (**Q**) we first need the following result.

6.3.1 Proposition

$(X,q) \in \mathbf{Q} \Rightarrow d\Phi$ is a **Q**-Cauchy structure

Proof: Let $\{\underline{F}_i \mid i \in I\}$ be an equivalence class in $d\Phi$ and suppose $\underline{F}$ is a filter on X satisfying the following condition:

$$\underline{U} \in U(X), \qquad \underline{U} \supset \underline{F} \Rightarrow \underline{U} \in \{\underline{F}_i \mid i \in I\}$$

We first prove that $\underline{F} \in d\Phi$. In view of the fact that $d\Phi$ is compatible with (X,q) only two cases can occur: Either there exists an $x \in X$ such that $\underline{F}_i$ q-converges to x for every $i \in I$, or $\alpha_q \underline{F}_i = \emptyset$ for every $i \in I$. In the first case, since (X,q) is pseudotopological we clearly have that $\underline{F}$ also q-converges to x, and consequently $\underline{F} \in d\Phi$. In the second case we clearly have $\alpha_q \underline{F} = \emptyset$. Moreover, for every $i \in I$, we have $\underline{F}_i \in c\Phi$ and all filters $\underline{F}_i$ are equivalent for the structure $c\Phi$. Since $c\Phi$ is itself pseudotopological, it follows that $\underline{F} \in c\Phi$. Consequently, $\underline{F} \in d\Phi$. Finally, in view of Proposition 1.4.9 we can conclude that $d\Phi$ is a Q-Cauchy space.

Next we come to our main theorem, which in view of Proposition 6.1.3 is a solution of problem (Q).

6.3.2 Theorem The following properties are equivalent:

1. There exists a compatible Q-Cauchy structure $\underline{C}$ on X such that $\Gamma(X,\underline{C}) = \Phi$.
2. (i) $(X,q) \in$ Q.
   (ii) Φ satisfies (D).
3. (i) $(X,q) \in$ Q.
   (ii) Φ is composition closed and determines the same initial topology as $\Gamma(X,d\Phi)$.

Proof: (1) $\Rightarrow$ (2) (i) If $\underline{C}$ is a compatible Q-Cauchy structure satisfying $\Gamma(X,\underline{C}) = \Phi$, let Y be a completion belonging to Q which is an R-extension. Then (X,q) is a subspace of Y and therefore it also belongs to **Q**. The proof of (ii) is identical to that of the implication (2) $\Rightarrow$ (3) in Theorem 6.2.6.

(2) $\Rightarrow$ (3) is exactly the implication (3) $\Rightarrow$ (4) in Theorem 6.2.6.

(3) $\Rightarrow$ (1) We first repeat the argument of (4) $\Rightarrow$ (1) in Theorem 6.2.6 to conclude that $\Gamma(X, d\Phi) = \Phi$. Then by Proposition 6.2.3(3) we know that $d\Phi$ is compatible with (X,q), and finally, by Proposition 6.3.1, $d\Phi$ is a **Q**-Cauchy structure.

6.4 EXTENSIONS IN THE CLASS OF ALL TOPOLOGICAL SPACES

In this section (X,q) is a topological space and Φ is a function class on X related to (X,q) by

$$\Phi \subset C(X,q)$$

Moreover, T is the class of all topological spaces. From Theorem 6.2.6 we immediately have the following.

6.4.1 Theorem Condition (D) is necessary for problem (T).

We give an example to show that (D) is not sufficient.

6.4.2 Example For $n \geq 1$ and for ω the first infinite ordinal, let I_n and I_ω be copies of [0,1]. Moreover let x_n and x_ω be copies of $x \in [0,1]$. We put $X = \cup_{n\geq 1} I_n \cup I_\omega$ and make it a space in the following way: For a point in a level I_n we take the usual neighborhoods as a neighborhood base. For a point $x_\omega \in I_\omega$ the sets

$$V_n^\varepsilon = \{x_\omega\} \cup \bigcup_{k\geq n} (x_k - \varepsilon, x_k + \varepsilon)$$

where $\varepsilon > 0$ and $n \geq 1$, form a neighborhood base.

The structure thus defined on X clearly is a Hausdorff topology. Next we show that it is Hausdorff closed. For this purpose we introduce the following auxilliary space: Let Y = $I_0 \cup I_\omega$, where I_0 again is another copy of [0,1]. For points on I_0 the usual neighborhoods form a base. For $x_\omega \in I_\omega$ the sets

$$\{x_\omega\} \cup (x_0 - \varepsilon, x_0 + \varepsilon) \qquad \varepsilon > 0$$

are basic neighborhoods.

Clearly every filter with an open base on Y has an adherence point in Y. Now let $\underline{G}$ be a filter with an open base on X. Then $\underline{G}$ has a trace on $\cup_{k\geq 1} I_k$. Moreover, the finer filter sup($\underline{G}$, stack $\cup_{k\geq 1} I_k$) again has an open base. So to prove that $\underline{G}$ has an adherence point, and without loss of generality, we may suppose that $\cup_{k\geq 1} I_k \in \underline{G}$.

Now the following two cases can occur: Either there is an index $n \geq 1$ such that $\underline{G}$ has a trace on $\cup_{k=1}^{n} I_k$ (and then clearly $\underline{G}$ has an adherence point), or we can find a $G_n \in \underline{G}$, for every $n \geq 1$, such that

$$G_n \cap \left(\bigcup_{k=1}^{n} I_k\right) = \emptyset$$

Let $\psi : X \to Y$ be the map which leaves I_ω pointwise fixed and which maps a point of I_n to the corresponding point of I_0. Clearly $\psi(\underline{G})$ is an open filter on Y. Let x_ω be the point in I_ω corresponding to the adherence point of $\psi(\underline{G})$ in Y.

We now claim that x_ω is an adherence point of $\underline{G}$ in X. Let V_i^ε be a basic neighborhood of x_ω, and let $G \in \underline{G}$. Then we have the following implications:

$$G' = G \cap G_i \Rightarrow \psi(G') \cap \psi(V_i) \neq \emptyset$$

$$\Rightarrow G' \cap V_i^{\varepsilon} \neq \emptyset$$

$$\Rightarrow G \cap V_i^{\varepsilon} \neq \emptyset$$

Now let $\underline{H}$ be the filter of finite complements on I_ω and let

$$\Phi = \{f \in C(X) \mid \text{stack } f(\underline{H}) \text{ converges}\}$$

Then clearly condition (D) is fulfilled. However, since X is Hausdorff closed, for every topological extension Y we have

$$C(Y)/X = C(X) \neq \Phi$$

To formulate sufficient conditions for the problem in the class T, we first define the following notation and prove the next lemma.

6.4.3 Definition For every filter $\underline{F}$ on (X,q) let

$$\underline{L}_q(\underline{F}) = \text{stack } \{0 \mid 0 \text{ q-open}, 0 \in \underline{F}\}$$

If no confusion can arise we simply write $\underline{L}(\underline{F})$.

6.4.4 Lemma For the following properties the implications (i) $\Rightarrow$ (ii) $\Rightarrow$ (iii) hold. If (X,q) is regular they are all equivalent.

(i) $d\Phi$ is topological.

(ii) $\alpha_q \underline{L}(\cap\{\underline{F} \in F(X) \mid \underline{F}\ t\Phi\text{-converges to } x, \alpha_q\underline{F} = \emptyset\}) = \emptyset$, $\forall x \in X$.

(iii) $\alpha_q \cap \{\underline{F} \in F(X) \mid \underline{F}\ t\Phi\text{-converges to } x, \alpha_q\underline{F} = \emptyset\} = \emptyset$, $\forall x \in X$.

Proof: (i) $\Rightarrow$ (ii) Suppose $d\Phi$ is topological and let $x \in X$. The filters in the collection $\{\underline{F} \in F(X) \mid \underline{F}$ $t\Phi$-converges to x, $\alpha_q\underline{F} = \emptyset\}$ are all equivalent in $(X,d\Phi)$. So $\underline{L}(\cap\{\underline{F} \in F(X) \mid \underline{F}$ $t\Phi$-converges to x, $\alpha_q\underline{F} = \emptyset\})$ belongs to $d\Phi$, and since it is not q-convergent, it has an empty q-adherence.

(ii) $\Rightarrow$ (iii) This is trivial.

(ii) $\Rightarrow$ (i) Suppose (ii) holds. We first show that (iii) implies that $d\Phi$ is pretopological. Let $\{\underline{F}_i \mid i \in I\}$ be an equivalence class for $d\Phi$. We consider again the two cases occurring in the proof of Proposition 6.3.1. In the first case, where all filters $\underline{F}_i$ converge to the same point, we have clearly that $\cap\underline{F}_i$ belongs to $d\Phi$.

The second case, where $\alpha_q\underline{F}_i = \emptyset$ for every $i \in I$, now has to be further divided into two subcases. Either $\alpha_{t\Phi}\,\underline{F}_i = \emptyset$ for every $i \in I$, and then $\cap\underline{F}_i$ is a Φ filter for which the $t\Phi$-adherence, and consequently also the q-adherence, are empty (so clearly then $\cap\underline{F}_i \in d\Phi$), or there is an $x \in X$ such that $\underline{F}_i$ $t\Phi$-converges to x for every $i \in I$. From the assumption (iii) we then can conclude that $\cap\underline{F}_i$ has an empty q-adherence, which then again implies $\cap\underline{F}_i \in d\Phi$. It now follows easily that in each of the cases, the assumption (ii) moreover implies that $\underline{L}(\cap\underline{F}_i)$ belongs to $d\Phi$.

Next we make the extra assumption that (X,q) is regular. Then if $x \in X$ and if $\underline{V}_q(y)$ stands for the neighborhood filter of a point y of X, the following implications hold:

(iii) $\Rightarrow$ sup $(cl_q\underline{V}_q(x), \cap\{\underline{F} \in F(X) \mid \underline{F}$ $t\Phi$-converges to x, $\alpha_q\underline{F} = \emptyset\})$ does not exist

$\Rightarrow$ sup $(\underline{V}_q(x), \underline{L}(\cap\{\underline{F} \in F(X) \mid \underline{F}$ $t\Phi$-converges to x, $\alpha_q\underline{F} = \emptyset\})$ does not exist

$\Rightarrow$ (ii)

In view of Proposition 6.1.4, in the next result, a sufficient condition is formulated for problem (T).

6.4.5 Theorem If Φ satisfies

(i) condition (D)

(ii) $\alpha_q \underline{L}(\cap\{\underline{F} \in F(X) \mid \underline{F}\ t\Phi\text{-converges to } x,\ \alpha_q \underline{F} = \emptyset\}) = \emptyset$, $\forall x \in X$,

then there exists a compatible T Cauchy structure $\underline{C}$ on (X,q) such that $\Gamma(X,\underline{C}) = \Phi$.

Proof: If Φ satisfies (i) and (ii), then, as in Theorem 6.2.6 it follows that $d\Phi$ is compatible and satisfies $\Gamma(X,d\Phi) = \Phi$. Moreover, in view of Lemma 6.4.4, $d\Phi$ is topological. Finally, from Proposition 1.4.9 it follows that $d\Phi$ is a T Cauchy structure.

However, condition (6.4.5)(ii) is not necessary for problem (T). This is shown by the following example.

6.4.6 Example We take one of the classical examples of a Hausdorff closed space which is not compact. In particular, let

$$X = (I \times \mathsf{N}) \cup \{\alpha\}$$

where $I = [0,1]$ and $\alpha \notin I \times \mathsf{N}$. For points in $I \times \mathsf{N}$ the neighborhoods of the product topology form a neighborhood base. The point α has a neighborhood base

$$\{V_n \mid n \in \mathsf{N}\}$$

where

$$V_n = (0,1] \times \{k \in \mathsf{N} \mid k \geq n\} \cup \{\alpha\}$$

Let $\Phi = C(X)$. Then clearly

$$\underline{C} = \{\underline{F} \in F(X) \mid \underline{F} \text{ converges}\}$$

is a compatible T Cauchy structure on X such that $\Gamma(X,\underline{C}) = \Phi$.

We now show that condition (ii) in Theorem 6.4.5 is not fulfilled. Let $\underline{F}$ be the filter on X generated by

$$\{ F_n \mid n \in \mathbb{N} \}$$

where

$$F_n = \{0\} \times \{k \in \mathbb{N} \mid k \geq n\}$$

For every $f \in C(X)$ and $\varepsilon > 0$, there is a neighborhood V_n of α such that

$$f(V_n) \subset (f(\alpha) - \frac{\varepsilon}{2}, f(\alpha) + \frac{\varepsilon}{2})$$

Using the continuity of f on each level, we can conclude that

$$f(F_n) \subset (f(\alpha) - \varepsilon, f(\alpha) + \varepsilon)$$

It follows that $\underline{F}$ $t\Phi$-converges to α. Moreover it is clear that $\alpha\underline{F} = \emptyset$. However, no filter with an open base and coarser than $\underline{F}$ has an empty adherence.

If the space (X,q) is locally compact, then the situation is different, as follows from the next result.

6.4.7 Proposition If (X,q) is locally compact, then for every $x \in X$ we have

$$\alpha_q \underline{L}(\cap\{ \underline{F} \in F(X) \mid \underline{F} \text{ } t\Phi\text{-converges to } x, \alpha_q\underline{F} = \emptyset\}) = \emptyset$$

Proof: In view of the regularity of the space, by application of Lemma 6.4.4, it suffices to prove that for every $x \in X$ the filter

$$\underline{H} = \cap\{ \underline{F} \in F(X) \mid \underline{F} \text{ } t\Phi\text{-converges to } x, \alpha_q\underline{F} = \emptyset\}$$

has an empty adherence.

Suppose, on the contrary, that $\underline{W}$ is a q-convergent ultrafilter, finer than $\underline{H}$. In view of the local compactness, $\underline{W}$ contains a compact set K. Moreover, we clearly have

$$\underline{W} \supset \cap \{\underline{U} \in U(X) \mid \underline{U}\ t\Phi\text{-converges to } x,\ \alpha_q \underline{U} = \emptyset\}$$

By Proposition 2 of [69] it follows that some ultrafilter in the collection on the right has to contain K. This, however, is impossible.

Combining Theorem 6.4.1, Theorem 6.4.5, and Proposition 6.4.6, we now immediately have the next result, which is a solution to problem (T) in the particular case of a locally compact space (X,q).

6.4.8 Corollary If (X,q) is locally compact, then the following properties are equivalent:

(i) Φ satisfies (D).

(ii) There exists a compatible T Cauchy structure $\underline{C}$ on (X,q) such that $\Gamma(X,\underline{C}) = \Phi$.

6.5 A COMPARISON OF CONDITIONS (B), (C), AND (D)

Again we consider the situation in the previous section. So (X,q) is a topological space and Φ is a function class on X related to (X,q) by

$$\Phi \subset C(X,q)$$

Further, T again stands for the class of all topological spaces. Condition (D), defined in 6.2.5, will be compared to the conditions (B) and (C), which were defined by A. Császár in [33] and formulated at the beginning of this chapter. (B) and (D) are necessary conditions, whereas (C) is sufficient for problem (T). Clearly the following implications hold:

$$\begin{array}{ccc} C & & \\ \Downarrow & \searrow & \\ B & \Rightarrow & D \end{array}$$

In general, however, no other implications are valid. Example 4.4 in [33] shows that (B) $\not\Rightarrow$ (C). Next we show that (D) $\not\Rightarrow$ (B).

6.5.1 Example We again take the space X constructed in Example 6.4.2, with the same function class Φ on it. We show that Φ does not satisfy condition (B).

Let $f : X \to R$ be defined as follows:

$$f(x_n) = x \qquad \text{for every } n \geq 1 \text{ and } x \in [0,1]$$

$$f(x_\omega) = x$$

Clearly then f is continuous. Moreover, since X is Hausdorff-closed, stack $f(\underline{G})$ converges whenever $\underline{G}$ is an open Φ-filterbase with empty q-adherence. However, $f \notin \Phi$.

For a regular space or for a locally compact space the situation is different. This will be shown in the following results.

6.5.2 Proposition On a regular topological space, conditions (B) and (D) are equivalent.

Proof: We have only to prove that (D) $\Rightarrow$ (B). Suppose Φ satisfies condition (D). Moreover, suppose $f \in C(X,q)$ and stack $f(\underline{F})$ converges whenever $\underline{F}$ is an open Φ-filter with empty q-adherence. Further, let $\underline{H}$ be an arbitrary Φ-filter with empty q-adherence. Then we have

$$\underline{L}_{t\Phi}\, \underline{H} \subset \underline{L}_q \underline{H} \subset \underline{H}$$

Consequently $\underline{L}_q \underline{H}$ again is a Φ-filter. Using the regularity of (X,q), we can conclude that $\underline{L}_q \underline{H}$ also has an empty q-adherence. It follows that stack $f(\underline{L}_q \underline{H})$ converges and then stack $f(\underline{H})$ converges too. Applying (D) we can conclude that $f \in \Phi$.

6.5.3 Proposition On a locally compact topological space conditions (B), (C), and (D) are equivalent.

Proof: Conditions (C) and (D) are necessary and sufficient for problem (T) when (X,q) is locally compact. For (C) this statement is exactly Corollary 4.6 in [33]. For (D) this was shown in Corollary 6.4.8. In view of Proposition 6.5.2 we can conclude that all three properties are equivalent.

6.6 EXTENSIONS IN THE CLASS OF ALL c-EMBEDDED SPACES AND IN THE CLASS OF ALL RELATIVELY ω-REGULAR SPACES

In this section (X,q) is a convergence space and Φ is a function class on X related to (X,q) by the assumption

$$\Phi \subset C(X,q)$$

Let Q be the class of all c-embedded convergence spaces and let S be the class of all relatively ω-regular convergence spaces defined in the following way: If (X,q) is a convergence space and $A \subset X$ then (X,q) is relatively ω-regular with respect to A if we have

$\underline{F}$ converges and $\sup(cl_{\omega q}\underline{F}$, stack A) exists $\Rightarrow$ $\sup(cl_{\omega q}\underline{F}$, stack A) converges

To solve problems (Q) and (S) we introduce the following new structure.

6.6.1 Definition

$e\Phi = \{\underline{F} \in F(X) \mid \underline{F}$ is q-convergent or $\underline{F} \in c\Phi,\ \alpha_{t\Phi}\underline{F} = \emptyset\}$

The next result follows immediately upon applying Lemma 6.2.1 and Proposition 6.2.3.

6.6.2 Proposition The following properties hold:

1. $e\Phi$ is a Cauchy structure on X.
2. $d\Phi \leq e\Phi$.
3. $e\Phi$ is compatible with (X,q).

We recall the definition of condition (C) introduced by Császár in [33].

6.6.3 Definition

If $f \in C(X,q)$ and stack $f(\underline{F})$ converges whenever $\underline{F}$ is an $u\Phi$ round Φ-filter with empty q-adherence, then $f \in \Phi$. (C)

For the solutions of problems (Q) and (S) we shall also need the following conditions on Φ.

6.6.4 Definition A function class Φ on (X,q) is called <u>regular</u> if $cl_{t\Phi}\underline{F}$ q-converges to x.

We shall need the following preliminary results.

6.6.5 Proposition The following implications hold:

1. (X,q) pseudotopological $\Rightarrow$ $e\Phi$ pseudotopological.
2. Φ regular $\Rightarrow$ $e\Phi$ μ-regular.
3. Φ point separating $\Rightarrow$ $e\Phi$ Cauchy separated.
4. $\Gamma(X,e\Phi) = \Phi \iff \Phi$ satisfies (C).

Proof: (1) This follows immediately from the facts that (X,q) is pseudotopological and $c\Phi$ is uniformizable.

(2) Suppose Φ is regular and let $\underline{F} \in e\Phi$. If $\underline{F}$ q-converges, then so does $cl_{t\Phi}\underline{F}$. Moreover, the inclusion $\Phi \subset \Gamma(X,c\Phi) \subset \Gamma(X,e\Phi)$ clearly always holds. So we can conclude that $cl_{t\Gamma(X,e\Phi)}\underline{F}$ also q-converges. If $\underline{F}$ is a Φ-filter with $\alpha_{t\Phi}\underline{F} = \emptyset$, then $cl_{t\Phi}\underline{F}$ again is a Φ-filter and then so is $cl_{t\Gamma(X,e\Phi)}\underline{F}$. Moreover, we clearly have

$$\alpha_{t\Phi}\, cl_{t\Gamma(X,e\Phi)}\underline{F} \subset \alpha_{t\Phi}\, cl_{t\Phi}\underline{F} = \alpha_{t\Phi}\underline{F} = \emptyset$$

(3) Suppose Φ is point separating. Then $t\Phi$ is Hausdorff. This implies that if $\underline{F}$ and $\underline{G}$ are not equivalent filters in $e\Phi$ then they are also nonequivalent in $c\Phi$. The rest of the proof then immediately follows from the fact that $c\Phi$ is Cauchy separated and from the inclusion $\Gamma(X,c\Phi) \subset \Gamma(X,e\Phi)$.

(4) The equivalence of $\Gamma(X,e\Phi) = \Phi$ and (C) will follow easily from the equality

$$e\Phi = \{\underline{F} \in F(X) \mid \underline{F} \text{ q-converges or } \exists\, \underline{G} \in c\Phi$$
$$\underline{G}\ u\Phi\text{-round},\ \alpha_q\underline{G} = \emptyset,\ \underline{G} \subset \underline{F}\}$$

which we prove now.

If $\underline{F}$ is a Φ-filter and $\alpha_{t\Phi}\underline{F} = \emptyset$, let $\underline{M}$ be the minimum element of the class of $\underline{F}$ in $c\Phi$. Then clearly $\underline{M}$ is $u\Phi$-round, $\alpha_q\underline{M} = \emptyset$, and $\underline{M} \subset \underline{F}$. To prove the other inclusion, let

$\underline{G} \in c\Phi$ be $u\Phi$-round and such that $\alpha_q\underline{G} = \emptyset$. Then clearly $\underline{G}$ cannot be the $t\Phi$ neighborhood filter of some point x. It follows that $\alpha_{t\Phi}\underline{G} = \emptyset$ and so $\underline{G} \in e\Phi$.

We can now formulate our solutions of problems (Q) and (S).

6.6.6 Theorem The following properties are equivalent:

(1) There exists a compatible Q-Cauchy structure $\underline{C}$ on X such that $\Gamma(X,\underline{C}) = \Phi$.
(2) (X,q) is pseudotopological, and Φ is regular, point separating, and satisfies (C).
(3) (X,q) is pseudotopological, Φ is regular, point separating, composition closed, and Φ and $\Gamma(X,e\Phi)$ have the same initial topology.

Proof: (1) $\Rightarrow$ (2) If $\underline{C}$ is a compatible Q-Cauchy structure on X with $\Gamma(X,\underline{C}) = \Phi$, let Y be a c-embedded completion of $(X,\underline{C})$. From Theorem 2.4.4 it follows that $\underline{C}$ is c^embedded, and then it is pseudotopological, μ-regular, and Cauchy separated. So we immediately have that (X,q) is pseudotopological, and that Φ is regular and point separating. To prove that Φ satisfies (C) it is sufficient, in view of Proposition 6.6.5(4) to show that $e\Phi \leq \underline{C}$. For this purpose, now suppose $\underline{F} \in \underline{C}$. From Proposition 3.5.3, it follows that $\underline{F} \in c\Phi$. Moreover, the following implications hold:

$$x \in \alpha_{t\Phi}\underline{F} \Rightarrow \dot{x} \geq cl_{t\Phi}\underline{F}$$

$$\Rightarrow \dot{x} \geq cl_{t\Gamma(X,\underline{C})}\underline{F}$$

$$\Rightarrow cl_{t\Gamma(X,\underline{C})}\underline{F} \text{ q-converges to } x$$

$$\Rightarrow \underline{F} \text{ q-converges to } x.$$

It follows that $\underline{F} \in e\Phi$.

(2) $\Rightarrow$ (3) This follows at once from Proposition 6.6.5(4) and Proposition 6.6.2.

(3) $\Rightarrow$ (2) In view of Proposition 6.6.5(4) it suffices to show that $\Gamma(X,e\Phi) = \Phi$. Clearly, in view of Proposition 6.6.2, we have $\Gamma(X,c\Phi) \subset \Gamma(X,e\Phi)$. To prove the other inclusion, let $f \in \Gamma(X,e\Phi)$ and let $\underline{F} \in c\Phi$. Either $\underline{F}$ $t\Phi$-converges or its $t\Phi$-adherence is empty. In the first case, we can conclude that stack $f(\underline{F})$ converges, using the $t\Phi$-continuity of f, which follows from the equality $t\Phi = t\Gamma(X,e\Phi)$. In the second case we have $\underline{F} \in e\Phi$ and then stack $f(\underline{F})$ also converges.

Finally, we have shown that $f \in \Gamma(X,c\Phi)$ and from the arbitrariness of f we can conclude that

$$\Gamma(X,c\Phi) = \Gamma(X,e\Phi)$$

Applying Proposition 5.1.2 we can conclude that

$$\Gamma(X,e\Phi) = \Phi$$

(2) $\Rightarrow$ (1) In view of Proposition 6.6.5, the assumptions made in (2) imply that $e\Phi$ is pseudotopological, μ-regular, Cauchy separated, and satisfies $\Gamma(X,e\Phi) = \Phi$. Moreover, from Proposition 6.6.2 we have that $e\Phi$ is compatible with (X,q). It follows from the results about the natural completion in Propositions 1.4.15 and 1.4.16 that $e\Phi$ is a **Q**-Cauchy structure.

From the proof of (1) $\Rightarrow$ (2) in the previous theorem and from Proposition 6.6.5 we now have the following result.

6.6.7 Corollary If there exists a compatible **Q**-Cauchy structure $\underline{C}$ on (X,q) with $\Gamma(X,\underline{C}) = \Phi$ then $e\Phi$ is the coarsest Cauchy structure with these properties.

6.6.8 Theorem The following properties are equivalent:

(1) There exists a compatible S-Cauchy structure $\underline{C}$ on X such that $\Gamma(X,\underline{C}) = \Phi$.
(2) Φ is regular and satisfies (C).
(3) Φ is regular, composition closed, and Φ and $\Gamma(X,e\Phi)$ have the same initial topology.

Proof: (1) $\Rightarrow$ (2) If $\underline{C}$ is a compatible S-Cauchy structure on X with $\Gamma(X,\underline{C}) = \Phi$ let Y be a relatively ω-regular completion which is an R-extension. We first show that $\underline{C}$ is μ-regular. Let $\underline{F} \in \underline{C}$. The filter stack $\underline{F}$ generated on Y converges in Y. For this filter we have

$$\sup(\mathrm{cl}_{\omega Y}\underline{F},\ \mathrm{stack}\ X) = \mathrm{cl}_{t\Phi}\underline{F}$$

It follows that $\mathrm{cl}_{t\Phi}\underline{F} \in \underline{C}$. The fact that Φ is regular and satisfies (C) now follows as in Theorem 6.6.6, (1) $\Rightarrow$ (2).

(2) $\Rightarrow$ (3) and (3) $\Rightarrow$ (2) This is completely analogous to the implication (2) $\Rightarrow$ (3) and (3) $\Rightarrow$ (2) in Theorem 6.6.6.

(2) $\Rightarrow$ (1) In view of Proposition 6.6.5, the assumptions made in (2) imply that $e\Phi$ is μ-regular and satisfies $\Gamma(X,e\Phi) = \Phi$. Moreover, from Proposition 6.6.2 we have that $e\Phi$ is compatible with (X,q). Now let Y be any strict completion of $(X,e\Phi)$. We show that it belongs to S. Let Θ be a filter on Y converging to y and suppose $\sup(\mathrm{cl}_{\omega Y}\Theta,\ \mathrm{stack}\ X)$ exists. Take a filter $\underline{F}$ containing X, converging to y in Y, and such that $\mathrm{cl}_Y\underline{F} \subset \Theta$. Then from the μ-regularity of $e\Phi$, we now have $\mathrm{cl}_{t\Phi}\underline{F} \in e\Phi$. Moreover, since

$$\mathrm{cl}_{t\Phi}\underline{F} = \sup(\mathrm{cl}_{\omega Y}\underline{F},\ \mathrm{stack}_Y X)$$

we also have

$$\mathrm{cl}_{t\Phi}\underline{F} \subset \sup(\mathrm{cl}_Y\Theta,\ \mathrm{stack}_Y X)$$

It follows that $\sup(cl_Y\Theta, stack_Y X)$ converges in Y. We finally can conclude that $e\Phi$ is an S-Cauchy structure.

In the next result we investigate how the previous theorem is related to our earlier work in [87]. We show that $e\Phi$ is related to the structure E_Φ introduced in [87].

6.6.9 Proposition If Φ is regular or point separating then we have

$$e\Phi = \{\underline{F} \in F(X) \mid \underline{F} \in c\Phi,\ \forall x \in X : \underline{F}\ t\Phi\text{-converges to } x \Rightarrow \underline{F}\ q\text{-converges to } x\}$$

Proof: Suppose $\underline{F} \in e\Phi$ and $t\Phi$-converges to x. It follows that $\underline{F}$ q-converges to some point $y \in X$. Using the regularity of Φ we have that $cl_{t\Phi}\underline{F}$ q-converges to y. Since $\dot{x} \geq cl_{t\Phi}\underline{F}$ we can now conclude that $\underline{F}$ q-converges to x. Using the point separability of Φ and the fact that $\underline{F}$ $t\Phi$-converges to both x and y we have again that $x = y$. The proof of the other inclusion is straightforward.

6.7 EXTENSIONS IN THE CLASS OF ALL RELATIVELY ω-REGULAR TOPOLOGICAL SPACES

In this section (X,q) is a <u>topological space</u>, Φ is a function class on X, and again we assume

$$\Phi \subset C(X,q)$$

As a corollary of the results of Section 6 we shall obtain results for problem (T), where T is the <u>class of relatively ω-regular topological spaces</u>. We first need the following result.

6.7.1 Proposition If (X,q) is topological then $e\Phi$ is a topological Cauchy structure.

Proof: Every equivalence class of $e\Phi$-Cauchy filters clearly has a minimum element, which is either a neighborhood filter $\underline{V}_q(x)$ for some $x \in X$, or a $c\Phi$-minimal Cauchy filter with empty $t\Phi$-adherence. In both cases it clearly has a q-open base.

6.7.2 Theorem The following properties are equivalent:

(1) There exists a compatible T-Cauchy structure $\underline{C}$ on (X,q) such that $\Gamma(X,\underline{C}) = \Phi$.

(2) Φ is regular and satisfies (C).

Proof: (1) $\Rightarrow$ (2) This follows immediately from Theorem 6.6.8, since every T Cauchy structure is an S Cauchy structure in the sense of Section 6.

(2) $\Rightarrow$ (1) If Φ is regular and satisfies (C) then it was shown in Theorem 6.6.8, (2) $\Rightarrow$ (1), that any strict completion Y of $(X,e\Phi)$ belongs to S. If we now take for Y the Kowalsky completion, then in view of Proposition 6.7.1 and the fact that this completion preserves the topological character, the space Y belongs to T.

6.7.3 Remark Solutions for the class of c-embedded topological spaces cannot be derived as a corollary of Theorem 6.6.6 in the same way as the solution for T in Theorem 6.7.2 was derived from Theorem 6.6.8. The reason is that the fact that $e\Phi$ is topological does not guarantee that the natural completion is topological.

References

1. J. Aarts. Cocompactifications. Indag. Math. 32 (1970) 9-21.

2. G. Artico, A Le Donne, and R. Moresco. Algebras of real-valued Uniform maps. Rend. Sem. Mat. Univ. Padova 61 (1979) 405-418.

3. R. Ball. Distributive Cauchy lattices. Algebra Universalis 18 (1984) 134-174.

4. R. Ball. Convergence and Cauchy structures on lattice ordered groups. Trans. Amer. Math. Soc. 259 (1980) 357-392.

5. B. Banaschewski. On Wallman's method of compactification. Math. Nachr. 27 (1963) 105-114.

6. B. Banaschewski: Extensions of topological spaces. Can. Math. Bull. 7 (1964) 1-22.

7. H. Bentley. Nearness spaces and extensions of topological spaces. Studies in Topology, Acad. Press (1975) 47-66.

8. H. Bentley, M. Hastings, and R. Ori. Rings of uniformly continuous functions, Categorical Topology, Sigma Series in Pure Math 5 Heldermaun Verlap (1984) 46-70.

9. H. Bentley and H. Herrlich. Extensions of topological spaces. Proc. Memphis Top. Conf. (1975) 129-184.

10. H. Bentley and H. Herrlich. The reals and the reals. Gen. Topol. Appl. 9 (1978) 221-232.

11. H. Bentley and H. Herrlich. Completion as reflection. Comm. Math. Univ. Car. 19 (1978) 541-568.

12. H. Bentley and H. Herrlich. Completeness for nearness spaces. Topological structures II, part 1, Math. Centre tracts 115 (1979), 29-40.

13. H. Bentley, H. Herrlich, and E. Lowen-Colebunders. The category of Cauchy spaces is cartesian closed. Topol. Appl. 27 2 (1987) 105-112.

14. H. Bentely, H. Herrlich, and E. Lowen-Colebunders. Convergence, (to appear).

15. H. Bentley, H. Herrlich, and W. Robertson. Convenient categories for topologists. Comm. Math. Univ. Car. 17 (1976) 207-227.

16. H. Bentley and B. Taylor. Wallman rings. Pacific J. Math. 58 (1975) 15-35.

17. H. Bentley and B. Taylor. On generalizations of C* embeddings for Wallman rings. J. Austral. Math. Soc. 25 (1978) 215-229.

18. H. Bentley and B. Taylor. The Stone Weierstrass theorem for Wallman rings. J. Austral. Math. Soc. 25 (1978) 230-240.

19. C. Biles. Gelfand and Wallman-type compactifications. Pacific J. Math. 35 (1970) 207-218.

20. C. Biles. Wallman-type compactifications. Proc. Amer. Math. Soc. 25 (1970) 363-368.

21. E. Binz. Continuous convergence on C(X). Lecture notes in Math. 469, Springer-Verlag (Berlin 1975).

22. E. Binz and H. Keller. Funktionenraüme in der Kategorie der Limesraüme. Ann. Acad. Sci. Fenn. Sec. A.I. 383 (1969) 1-21.

23. G. Birkhoff. Lattice theory. Amer. Math. Soc. Colloquium publ. XXV (1967).

24. N. Bourbaki. Topologie générale. Eléments de Mathématique Fas. II Chap. 1,2 Hermann (Paris 1965).

25. G. Bourdaud. Espaces d'Antoine et Semi-espaces d'Antoine. Cahiers de Top. et Géom. Diff. 16 (1975) 107-133.

26. H. Butzman and B. Muller. Topological c-embedded spaces. Gen. Topol. Appl. 6 (1976) 17-20.

27. E. Čech. Topological spaces. Publishing house of the Czechoslovak Academy of Sciences (Prague), Interscience-John Wiley & Sons (1966).

28. G. Choquet. Convergences. Ann. Univ. Grenoble, Sect. Sci. Math. Phys. 23 (1948) 57-112.

29. C. Cook and H. Fisher. On equicontinuity and continuous convergence. Math. Ann. 159 (1965) 94-104.

30. C. Cook and H. Fisher. Uniform convergence structures. Math. Ann. 173 (1967) 290-306.

31. A. Császár. Function classes, compactifications, realcompactifications. Ann. Univ. Sci. Budapest Eötvös Sect. Math. 17 (1974) 139-156.

32. A. Császár. Some problems concerning C(X). General Topology and its relations to Analysis and Algebra IV. Lecture Notes in Math. 609 (1976) 41-55.

33. A. Császár. Filter closed function classes. Math. Struct.-Comp. Math.-Math. Mod. (1984) 131-135.

34. W. Feldman. Axioms of countability and the algebra C(X). Pacific J. Math. 47 (1973) 81-89.

35. H. Fisher. Limesraüme. Math. Ann. 137 (1959) 269-303.

36. R. Frič and D. Kent. On the natural completion functor for Cauchy spaces. Bull. Austral. Math. Soc. 18 (1978) 335-343.

37. R. Frič and D. Kent. Completion functors for Cauchy spaces. Internat. J. Math. and Math. Sci 2,4 (1979) 589-604.

38. R. Frič and D. Kent. A completion functor for Cauchy groups. Internat. J. Math. and Math. Sci. 4 (1981) 55-65.

39. R. Frič and D. Kent. A nice completion need not be strict. General topology and its relation to Analysis and Algebra V (1981) 189-192.

40. O. Frink. Compactifications and seminormal spaces. Amer. J. Math. 86 (1964) 602-607.

41. A. Frölicher and A. Kriegl. Convergence vector spaces for analysis. Convergence structures 1984 Proc. Conf. Bechyně. Akademie Verlag, (Berlin 1985) 115-125.

42. Z. Frolik. Concerning topological convergence of sets. Czechoslovak Math. J. 10 (1960) 168-180.

43. W. Gähler. Beiträge zur Theorie der Limesraüme. Theory of sets and topology (Berlin 1972) 161-197.

44. W. Gähler. On reflective and coreflective subcategories of

Lim, Lun and Lvec. General Topology and its relations to Analysis and Algebra IV (1976) 136-141.

45. W. Gähler. Grundstrukturen der Analysis I. (Berlin 1977), II (Berlin 1978).

46. S. Gähler, W. Gähler, and G. Kneiss. Completion of pseudo-topological vector spaces. Math. Nachr. 75 (1976) 185-206.

47. R. Gazik and D. Kent. Regular completions of Cauchy spaces via function algebras. Bull. Austral. Math. Soc. 11 (1974) 77-88.

48. L. Gillman and M. Jerison. Rings of continuous functions. Springer Verlag, 2nd ed. (Berlin 1976).

49. A. Hager. On inverse-closed subalgebras of C(X). Proc. London Math. Soc. 3 (1969) 233-257.

50. A. Hager and D. Johnson. A note on certain subalgebras of C(X). Canad. J. Math. 20 (1968) 389-393.

51. H. Hahn. Theorie der reellen functionen. (Berlin 1921).

52. D. Harris. Structures in Topology. Mem. Amer. Math. Soc. 115 (1971).

53. S. Hechler. On a notion of weak compactness in nonregular spaces. Studies in topology, Acad. Press (1975) 215-237.

54. M. Hendriksen and D. Johnson. On the structure of a class of archimedian lattice ordered algebras. Fund. Math. 50 (1961) 73-94.

55. H. Herrlich. A concept of nearness. Gen. Topol. Appl. 4 (1974) 191-212.

56. H. Herrlich. On the extendibility of continuous functions. Gen. Topol. Appl. 5 (1974) 213-215.

57. H. Herrlich. Topological structures I. Math. Centre tracts 52 (1974) 59-122.

58. H. Herrlich. Categorial topology 1971-1981. Mathematik Arbeitspapiere 24 (Bremen 1981).

59. E. Hewitt. On two problems of Urysohn. Ann. Math. 47 (1946) 503-509.

60. E. Hewitt. Certain generalizations of the Weierstrass approximation theorem. Duke Math. J. 14 (1947) 419-427.

61. I. Isbell. Algebras of uniformly continuous functions. Ann. Math. 68 (1958) 96-125.

62. A. Ivanov. Regular extensions of topological spaces. Contr. extension theory symp. (Berlin 1967).

63. M. Katĕtov. On H closed extensions of topological spaces. Cas. pes. mat. fys. roc. 72 (1947) 17-31.

64. M. Katĕtov. On continuity structures and spaces of mappings. Comm. Math. Univ. Car. 6 (1965) 257-278.

65. M. Katĕtov. Convergence structures. General topology and its relation to Analysis and Algebra II (1966) 207-216.

66. H. Keller. Die Limes-uniformisierbarkeit der Limesraüme. Math. Ann. 176 (1968) 334-341.

67. D. Kent. Convergence functions and their related topologies. Fund. Math. 54 (1964) 125-133.

68. D. Kent. On the order topology in a lattice. Journ. of Math 10 (1966) 90-96.

69. D. Kent. Decisive convergence spaces. Math. Ann. 184 (1970) 215-223.

70. D. Kent. A note on regular Cauchy spaces. Pacific J. Math. 94 (1981) 333-339.

71. D. Kent, K. Mc Kennon, G. Richardson, and M. Schroder. Continuous convergence in C(X). Pacific J. Math. 52 (1974) 271-279.

72. D. Kent and G. Richardson. Regular completions of Cauchy spaces. Pacific J. Math. 51 (1974) 483-490.

73. D. Kent and G. Richardson. Compactifications of convergence spaces. Internat. J. Math. and Math. Sci. 2,3 (1979) 345-368.

74. D. Kent and G. Richardson. Completely regular and ω-regular spaces. Proc. Amer. Math. Soc. 82 (1981) 649-652.

75. D. Kent and G. Richardson. Cauchy spaces and their completions. Abhandelungen der Akademie der Wissenschaften der DDR II (1984) 103-112.

76. D. Kent and G. Richardson. Cauchy spaces with regular completions. Pacific J. Math. 111 (1984) 105-116.

77. D. Kent, G. Richardson, and R. Gazik: T-regular closed convergence spaces. Proc. Amer. Math. Soc. 51 (1975) 461-468.

78. G. Kneiss. On regular completion of pseudo uniform spaces. Colloquia Math. Soc. Janos Bolyai 23 (1978).

79. H. Kowalsky. Limesraüme und Komplettierung. Math. Nachr. 12 (1954) 301-340.

80. R. Lee. The category of uniform convergence spaces is cartesian closed. Bull. Austral. Math. Soc. 15 (1976) 461-465.

81. E. Lowen-Colebunders. Completeness properties for convergence spaces. Pacific J. Math 70 (1977) 401-411.

82. E. Lowen-Colebunders. The Choquet hyperspace structure for convergence spaces. Math. Nachr. 95 (1980) 17-26.

83. E. Lowen-Colebunders. On the uniformization of the hyperspace of closed convergence. Math. Nachr. 105 (1982) 35-44.

84. E. Lowen-Colebunders. Uniformizable Cauchy spaces. Internat. J. Math. and Math. Sci. 5 (1982) 513-527.

85. E. Lowen-Colebunders. Algebras of Cauchy continuous maps. J. Math. Anal. Appl. 104 (1984) 408-417.

86. E. Lowen-Colebunders. On the regularity of the Kowalsky completion. Canad. J. Math. 36 (1984) 58-70.

87. E. Lowen-Colebunders. On composition closed function classes. Acta Math. Acad. Sci. Hung. 44 (1984) 181-189.

88. E. Lowen-Colebunders. Function classes of Cauchy continuous maps. Aggregatie-thesis, Vrije Universiteit Brussel (1986).

89. K. Mc Kennon. The strict topology and the Cauchy structure of the spectrum of a C*-algebra. Gen. Topol. Appl. 5 (1975) 249-262.

90. A. Machado. Espaces d'Antoine et pseudotopologies. Cahiers de Top. et Géom. Diff. 14 (1973) 309-327.

91. J. Mikusinski. Operational Calculus. Pergamon Press (New York 1959).

92. S. Mrówka. On the convergence of nets of sets. Fund. Math. 45 (1958) 237-246.

93. S. Mrówka. Some approximation theorems for rings of unbounded functions. Notices of Amer. Math. Soc. 11 (1964) 666.

94. S. Mrówka. On some approximation theorems. Nieuw archief voor wiskunde 16 (1968) 94-111.

95. B. Muller. T_3 completions of convergence vector spaces. General topology and its relation to Analysis and Algebra IV (1976) 198-307.

96. L. Nel. A categorical approach to topological character theory. Gen. Topol. Appl. 6 (1976) 241-258.

97. H. Poppe. Characterisierung der Kompactheit eines topologischen Raumes X durch konvergenz in C(X). Math. Nachr. 29 (1965) 205-216.

98. J. Porter and W. Feldman. Convergence structures for operators. Proc. of the conf. on convergence structures, Cameron University (1980) 113-118.

99. J. Ramaley and O. Wyler. Cauchy spaces I and II. Math. Ann. 187 (1970) 175-199.

100. E. Reed. Completion of uniform convergence spaces. Math. Ann. 194 (1971) 83-108.

101. G. Richardson. A Stone Čech compactification for limit spaces. Proc. Amer. Math. Soc. 25 (1970) 403-404.

102. G. Richardson and D. Kent. Regular compactifications of convergence spaces. Proc. Amer. Math. Soc. 11 (1972) 571-573.

103. G. Richardson and D. Kent. The regularity series of a convergence space. Bull. Austral. Math. Soc. 13 (1975) 21-44.

104. W. Robertson. Convergence as a nearness concept. Thesis Carleton University Ottawa (Ontario 1975).

105. N. Sanin. On special extensions of topological spaces. Dokl. Akad. Nauk SSSR 38 (1943) 6-9.

106. A. Sapsal and Z. Szabo. Pseudocompact extensions. Ann. Univ. Sci. Budapest Sect. Math. XXV (1982) 251-256.

107. M. Schroder. Solid convergence spaces. Bull. Austral. Math. Soc. 8 (1973) 443-459.

108. R. Snipes. Cauchy regular functions. J. Math. Anal. Appl. 79 (1981) 18-25.

109. F. Schwartz. Connections between convergence and nearness. Lect. Notes in Math. 719 (1979) 345-357.

110. F. Treves. Topological vector spaces, distributions and kernels. Acad. press (New York 1967).

111. E. Wagner. Convergence structures for Mikusinski operators. Proc. of the special session on convergence structures. University of Nevada (Reno 1980) 236-240.

112. H. Wallman. Lattices and topological spaces. Ann. Math. 39 (1938) 112-126.

113. A. Ward. On relations between certain intrinsic topologies in partially ordered sets. Proc. Cambridge Philos. Soc. 51 (1955) 254-261.

114. S. Willard. General topology. Addison Wesley Reading (Mass. 1970).

115. O. Wyler. Ein Komplettierung für uniforme Limesräume. Math. Nachr. 46 (1970) 1-12.

116. O. Wyler. On the categories of general topology and topological algebra. Arch. Math. 22 (1971) 7-17.

Index

RANDALL LIBRARY-UNCW
3 0490 0096503 $